The Joy of Science
Part I

Professor Robert M. Hazen

THE TEACHING COMPANY ®

PUBLISHED BY:

THE TEACHING COMPANY
4840 Westfields Boulevard, Suite 500
Chantilly, Virginia 20151-2299
1-800-TEACH-12
Fax—703-378-3819
www.teach12.com

Copyright © The Teaching Company, 2001

Printed in the United States of America

This book is in copyright. All rights reserved.

Without limiting the rights under copyright reserved above,
no part of this publication may be reproduced, stored in
or introduced into a retrieval system, or transmitted,
in any form, or by any means
(electronic, mechanical, photocopying, recording, or otherwise),
without the prior written permission of
The Teaching Company.

ISBN 1-56585-645-7

Robert M. Hazen, Ph.D.

Clarence Robinson Professor of Earth Science, George Mason University

Robert M. Hazen, research scientist at the Carnegie Institution of Washington's Geophysical Laboratory and Clarence Robinson Professor of Earth Science at George Mason University, received the B.S. and S.M. in geology at the Massachusetts Institute of Technology (1971), and the Ph.D. at Harvard University in earth science (1975). After studies as NATO Fellow at Cambridge University in England, he joined Carnegie's research effort in 1976.

Hazen has authored 220 articles and 15 books on science, history, and music. He received the Mineralogical Society of America Award (1982), the American Chemical Society Ipatieff Prize (1986), the ASCAP-Deems Taylor Award (1989), the Educational Press Association Award (1992), Fellowship in the American Association for the Advancement of Science (1994), and the American Crystallographic Association's Science Writing Award (1998). Hazen's research focuses on high-pressure organic synthesis and the origin of life.

Professor Hazen is active in presenting science to a general audience. His articles on science for a general readership have appeared in *Newsweek*, *The New York Times Magazine*, *Smithsonian Magazine*, *Technology Review*, and *Scientific American*. At George Mason University he has developed courses and companion texts on scientific literacy. His books include the best selling *Science Matters: Achieving Scientific Literacy* and *The Sciences: An Integrated Approach*. Hazen serves on advisory boards of the National Committee for Science Education, NOVA (WGBH TV Boston), *Encyclopedia Americana*, and the Carnegie Council. He appears frequently on radio and television programs on science.

Robert Hazen is also a professional trumpeter. He has performed with numerous ensembles, including the National Symphony Orchestra, the Boston Symphony Orchestra, and orchestras of the Metropolitan Opera, the New York City Opera, the Kirov Ballet, and the Royal Ballet. He is presently a member of the Washington Chamber Symphony and the National Gallery Orchestra.

Table of Contents
The Joy of Science
Part I

Professor Biography		i
Course Scope		1
Lecture One	The Nature of Science	2
Lecture Two	The Scientific Method	19
Lecture Three	The Ordered Universe	35
Lecture Four	Celestial and Terrestrial Mechanics	50
Lecture Five	Newton's Laws of Motion	67
Lecture Six	Universal Gravitation	82
Lecture Seven	The Nature of Energy	98
Lecture Eight	The First Law of Thermodynamics	113
Lecture Nine	The Second Law of Thermodynamics	129
Lecture Ten	Entropy	145
Lecture Eleven	Magnetism and Static Electricity	162
Lecture Twelve	Electricity	178
Concordance to the Science Content Standards of the *National Science Education Standards*		195
The Periodic Table of the Elements		199
Timeline		200
Glossary		205
Biographical Notes		213
Bibliography		218

Note to Video and DVD Customers: The transcripts for *The Joy of Science* were taken from the audio version of the course. Two versions were recorded so that Professor Hazen could better utilize the visual nature of the video course. The content of the two versions is the same, and the language is very similar, but video and DVD customers will find differences between the spoken and written word.

The Joy of Science
Part I

Scope:

Part I of our exploration of the central concepts of science begins with an examination of science as a way of knowing (Lectures One and Two), and then considers the overarching phenomena of forces and motions (Lectures Three through Six), energy (Lectures Seven through Ten), and magnetism and electricity (Lectures Eleven and Twelve). Of the many ways devised to understand our place in the cosmos, science holds a special place as a method designed to ask and answer questions about tangible aspects of the physical world. Scientific knowledge depends on independently verifiable data obtained from observations, experiments, and theoretical reasoning.

The scientific method, in its idealized form, is a cycle of observation, synthesis, hypothesis, and prediction that leads to more observations. This process developed gradually, coming to full flower in the 16th century with the work of Nicolas Copernicus and subsequent astronomers on the nature of celestial motions, as well as Galileo Galilei and his contemporaries on the laws of terrestrial motion. Isaac Newton synthesized these seemingly separate areas of research in his universal laws of motion and gravitation.

Energy, the ability to do work, is essential in every endeavor, yet its scientific description is difficult and abstract. Two laws of thermodynamics systematize the behavior of energy. The first law states that while energy occurs in many interchangeable forms, the total amount of energy is constant. The second law of thermodynamics states that energy tends to spread out, shifting from more useful to less useful forms.

Newton's laws established a systematic approach to the study of forces, such as magnetism and electricity, which were long thought to be unrelated phenomena. The invention of the battery, and subsequent studies of electrical phenomena, set the stage for the synthesis of electricity and magnetism.

Lecture One
The Nature of Science

Principle: *"Science is a way of knowing about the natural world, based on reproducible observations and experiments."*

Scope:

Of the many ways devised to understand our place in the cosmos, science holds a special place as a method designed to ask and answer questions about tangible aspects of the physical world. Scientific knowledge depends on independently verifiable data obtained from observations, experiments, and theoretical reasoning.

Modern society is profoundly affected by the discoveries of science and their applications through technology. Scientific literacy, which is important for every citizen, includes an understanding of the great principles that describe the workings of the physical world, as well as familiarity with the process of inquiry by which those principles were discovered.

Academic science is conveniently divided into four major disciplines: physics, chemistry, earth science, and biology, though the natural world does not recognize such disciplinary boundaries.

Outline

I. Humans have devised many ways of knowing: religion, philosophy, the arts, ethics, and science all can help us to understand our place in the natural world. Science holds a special place as a method designed to ask and answer questions about the tangible aspects of the physical world.
   A. Science is based on reproducible observations, controlled experiments, and theoretical reasoning. Science is fundamentally different from other ways of knowing because it is based on independently verifiable facts about physical phenomena.
      1. Science differs from religion, which depends on revealed truth.
      2. Science differs from the arts, which depends on each artist's unique vision.

3. Science differs from political science and social science, in which the richness of the discipline results from a multiplicity of interpretations of past and future events.
4. Science differs from pseudo-sciences, which are not based on reproducible and independently verifiable observations.

B. The process of science provides a powerful tool for observing the world, learning how it works, making predictions about future events, and discovering ways to modify our surroundings.

II. In the modern world, all citizens need to be scientifically literate.

A. Scientific literacy helps consumers make informed decisions. Science content provides a basis for understanding how and why products work; the scientific process of inquiry provides a framework for critical thinking about personal choices.

B. Scientific literacy is important in many jobs that depend on science, as well as on the technologies that are developed from scientific discoveries. Modern medicine, law, and business all depend on science.

C. Scientific literacy provides a foundation for teaching children about their world. Parents and teachers who are scientifically literate can reinforce a child's natural curiosity by exploring the natural world together.

D. Scientific literacy allows a person to share the richness of humanity's great ongoing adventure of discovery and exploration. Every day, scientists are discovering and sharing new things that no human ever knew before.

III. Everyone can achieve scientific literacy. The great principles of science can be presented without relying on complex vocabulary or mathematical abstraction.

A. Several contrasting definitions of scientific literacy have confused the issue and hindered science education reform. Scientific literacy is not technological literacy, nor is it the stepping stone exclusively for future scientists. You do not have to be a scientist to appreciate the great discoveries of science.

B. The National Science Education Standards represent a consensus of thousands of scientists and educators regarding the content and presentation necessary for all citizens to achieve scientific literacy. This document provides a sound building code for science curricula.
  1. Scientific literacy is based on understanding a few overarching principles, rather than extensive vocabulary and factual information.
  2. Scientific literacy is also based on understanding the scientific method—the process of inquiry by which questions about the natural world are asked and answered.

C. The objectives of this lecture series are to introduce the great principles of science—to explore how scientists discovered these sweeping ideas, and to show how the great principles come into play in our lives, in often surprising and wonderful ways.

IV. Science is a human endeavor, with its own distinctive organizational structures. The rigid division of science into separate departments may have hindered scientific literacy of nonscientists, but there is a rationale to this structure. In fact, there is a clear hierarchy of scientific disciplines.

  A. Physics, the study of matter in motion, is first in this hierarchy, because the laws of physics must apply to every natural system—large or small, living or inanimate.
  1. Classical physics deals with the everyday phenomena of matter, energy, forces, and motion.
  2. Modern physics deals with realms beyond our daily experience: very small, very massive, and very fast objects require new physical laws, as codified in the fields of quantum mechanics, nuclear physics, particle physics, and relativity.

  B. Chemistry is the study of atomic interactions, especially the study of chemical reactions and the formation of new materials.
  1. Modern chemistry is rooted in alchemical studies of the Middle Ages; it is an empirical science in which inspired guesswork often leads to major advances.

      2. Chemical research and chemical engineering are closely linked. It is often possible to scale up an experimental discovery made in a test tube to a full-scale manufacturing operation.
- **C.** Earth science is the study of the origin, present state, and dynamic processes of the Earth, as well as other nearby planets.
  1. Earth science is a young field that grew out of the practical experience of prospecting and mining.
  2. Geology is the study of rocks and soils and their distribution.
  3. Geophysics is the study of earth's dynamic processes, such as tides, hurricanes, and earthquakes.
- **D.** Biology is the study of living systems, which are by far the most complex objects we know. Because of its importance to health and medicine, biology is the largest and best funded of the sciences.
  1. Biology is an extremely diverse discipline because there are many ways to study living things.
  2. The most obvious applications of biology are in health and medicine.
- **E.** In spite of the formal academic separation of science disciplines into physics, chemistry, earth science, and biology, the natural world is integrated and knows no such boundaries.

Essential Reading:

National Research Council, *The National Science Education Standards*.

Trefil and Hazen, *The Sciences: An Integrated Approach*, Chapter 1.

Supplemental Reading:

Holton, *Thematic Origins of Scientific Thought*, Introduction and Chapter 1.

Questions to Consider:

1. In what ways has science affected your life in the past 24 hours?
2. If you were transported to a pre-science society, what would you miss most? In what ways might you benefit?

Lecture One—Transcript
The Nature of Science

Hi, I'm Robert Hazen, Clarence Robinson Professor of Earth Science at George Mason University in Fairfax, Virginia. For the last 10 years, I've worked with Jim Trefil at George Mason in developing courses that examine the basic principles of science and the many ways those principles come into play in your life.

I'm also a staff scientist at the Carnegie Institution of Washington Geophysical Laboratory. For many years there, my research has been focused on materials that form the Earth's deep interior. I've studied atomic structures of crystals at high pressure such as those pressures that occur deep within our planet. I've also begun an exciting project on the chemical processes that may have led to the origin of life and, especially, in the roles that minerals may have played in life's origins. I'm going to tell you more about some of this really exciting research in the coming lectures.

I have a passion as a teacher, and that passion is to share with you the joy of science, the astonishing discoveries, the mind-bending insights, and the transforming applications of science as well. See, science is the greatest ongoing human adventure, the greatest adventure of the human mind, and you can share in that adventure.

To begin, we need to examine what makes science special. We need to think about science as a way of knowing. So, I ask you, how do you know what you know? Humans have devised many different ways of knowing. We have religion, we have philosophy, there's the arts, there's ethics, and, of course, there's science. All these can help us understand our place in the natural world. But, of these very pursuits, science holds a special place; it's a method designed to ask and answer questions about the tangible aspects of the natural world. Indeed, science is a way of knowing about the natural world that's based on reproducible observations and carefully controlled experiments.

In this series of lectures on *The Joy of Science*, we're going to explore the full range of physical and life sciences. That includes topics that are traditionally allocated to things like physics, geology, astronomy, chemistry, and, of course, biology. We're going to find that these disciplines all share a common methodology and that, in fact, nature doesn't know anything about what we call these disciplinary boundaries. Throughout this series, we're going to see that the sciences are linked by a shared methodology and a common set of natural laws.

I have three objectives in this very first lecture, and these three objectives are first, to describe briefly the nature of science as a way of knowing and to see how science differs from other ways of knowing. Then, I want to explain something about the philosophy of this very unusual, integrated course. Why would I adopt this nontraditional approach to teaching science? Then third, I want to look briefly at the social structure of science, if you will, because science is a kind of human endeavor.

Science is a way of knowing. Science is based on reproducible observations. It's based on controlled experiments and on theoretical reasoning. One very simple and elegant definition of science, therefore, is it's the search for laws that describe the organization and the evolution of the universe. This means that science is fundamentally different from other ways of knowing, because it's based on independently verifiable facts about physical phenomena and it strives for a consensus based on those facts.

I have the great pleasure of participating in a course on visual thinking. That course is team taught by an artist, a psychologist, and myself, a scientist. We often begin that course by telling the story of the three umpires. The story goes like this: After a game, three umpires are sitting in a bar, and they're talking about how they call balls and strikes. The first umpire is an artist by day, and he tells his colleagues, "I calls 'em as I sees 'em." The second umpire, a scientist by day, shakes his head and says, "That's not the way to do it. I calls 'em as they really are." Then the third umpire says, "Naw, you're both wrong." The third umpire is a psychologist by day, and he says, "They ain't nothing 'til I calls 'em." The point is that there are lots of different ways of knowing.

I've defined science as a way of knowing based on reproducible observations, on controlled experiments, on mathematical reasoning. Think about how science, therefore, differs from other ways of knowing. Science differs from religion, because religion is based on revealed truth, sacred text, and other sources, and that has to be accepted on faith. The Bible, if it's taken to be literal truth, must be accepted on faith. You might also experience personal revelation or visions, and those have to be beyond questioning. Faith, if it's firm, has to supersede any observations that you can make in the physical world.

Science also differs from the arts. In the arts, truth is explored through each individual artist's unique vision. I want to tell you a story about the great 20th-century artist Pablo Picasso. This occurred late in his life, when he was at the height of his fame. It's said that Picasso was riding on a train in Europe. This was one of those trains where there are compartments, and people face each other in a small compartment in the train. As the story goes, one of Picasso's fellow travelers recognized him and started mumbling, pretty much so that everyone could hear but to no one in particular. He said, "It's a disgrace; it's a sham. I can't believe people fall for that stuff." After a while, Picasso interrupted and said, "Excuse me, what are you talking about?" He said, "Modern art. I don't understand it. I don't like it." "What don't you like about it?" Picasso says. "Well, it doesn't look like anything. It's not like reality." Picasso asks, "What is like reality?" The passenger thinks for a while, then says, "Ah, here." He pulls out his wallet. "Here's a picture of my wife. See, this is like reality, this really looks like her." Picasso pauses and says, "My goodness, she's awfully small and very flat." The point is that art is all a matter of your point of view. There are no absolute truths.

Science also differs from political science or from social science in a rather intriguing way. All of these so-called sciences are grounded in verifiable facts about the world around us. In science, the ultimate goal is to reach a consensus based on reproducible and verifiable facts. However, in politics and history and sociology and so forth, there are many equally valid interpretations of events. There are lots of plausible policies in interpretation, because people and their experiences are so richly varied. Indeed, the richness of these disciplines results from the multiplicity of interpretations of past, present, and future events.

Science also differs from astrology and psychic phenomena, UFOs, for example. These are called pseudosciences, and they're not based on reproducible and independently verifiable observations. I have to make a key point here: Most scientists don't reject the existence of UFOs. They don't say that psychic phenomena are impossible. It's just that reports of these phenomena are invariably sort of anecdotal. They're not subject to independent observations. They're not subject to verification. If a UFO lands and we can study it, then it may become science, but until that happens, there's not much a scientist can say one way or another. If and when reproducible evidence is available, then these subjects are going to satisfy the scientific test for knowledge.

The process of science provides a powerful tool for observing the way the world works, for making predictions about future events, for discovering ways to modify our surroundings. Imagine how vastly difficult a society without science would seem to us. Think about that. I think we'd immediately notice fundamental differences in our material well being. We'd lose technologies associated with health. We'd have much more limited kinds of food. We wouldn't have very good transportation. Communication would be virtually nonexistent. The housing would be primitive, and so much more.

Very soon, we'd notice more subtle, but equally profound, differences, such as differences in logic systems. For example, the belief in a predictable universe and the understanding of cause and effect. I suspect that, in the long run, it would be easier to adapt to the material hardships than to the loss of a rational framework of our society.

As we proceed in this 60-lecture survey, I hope that the unique character and the power and the joy of science will become clear to you. Before we proceed with the content of science, though, I owe you an explanation. Why is this course organized in such a nontraditional way? Jim Trefil and I developed this course in the joy of science in the early 1990s. It was in response to a growing call for a scientific literacy for all Americans. In the modern world, all citizens need to be scientifically literate. Any issue of any major newspaper or any news magazine is going to have numerous stories that are directly related to science.

You can do this yourself. Just look at this morning's newspaper. There are so many stories about science. On the front page, there is a story about pharmaceuticals, a story about scientists using embryos for research. There is another study about energy, a house study. And, of course, the weather always appears on the first page. That's just the first page. On the second page, there are stories about the FDA regulating certain fertilization procedures and a flood in West Virginia—more about science. On the next page, there is a big article about an embassy bomber, where much of the evidence in that trial was through forensics, through the science of chemical analysis. The next page, space shuttle launch, various medical treatments they're describing. The next page, we see a shark attack in Florida. And page after page, you see this—articles on the environment, articles on the drug war. Even on the editorial page—you might think that was all politics, but no. Look at this, an article on nuclear weapons in Russia, an article on environmental concerns, and an article on nuclear waste disposal—all in just the first section of today's paper. And you can do this for any day of the week.

Here's yesterday's paper: missile defense on the front page, "Fossil Suggests Earliest Human Ancestor," cloning embryos. Look at the day before: on the front page, evidence of toxins in the environment, warming shrinks Peruvian glaciers. And the day before—I can do this day after day—"Tougher Battles Bloom for Bush," regarding the environment, regarding medical issues; "SUVs Drive Area Pollution"; and cocaine, the drug wars again. Every single day, we find article after article about science.

It's clear that science pervades the news. Everyone needs to be scientifically literate. In fact, Jim Trefil and I cite four principal reasons for the importance of scientific literacy. The first reason is that scientific literacy helps consumers make informed decisions. Science content provides the basis for understanding how and why products work, and the scientific process of inquiry provides a framework for critical thinking about your personal choices. You have to make choices about health care: what medications to take, what treatments you're going to accept. You need the background to learn more from your doctor. We all make personal choices about diet and nutrition, and science informs those decisions. You need to understand the role of diet and exercise in your life and that, in today's world, we're increasingly confronted by urgent choices about the environment. These environmental issues face us all the time. You're asked to choose the correct kinds of detergents and diapers, automobiles. You're asked to recycle; you're asked to conserve. Science provides a convincing context for these efforts.

There's a second reason to become scientifically literate, and that's because of the workplace. So many of today's jobs depend directly or indirectly on science, as well as on technologies that are developed from scientific discoveries. Doctors and their patients need to understand the benefits and risks of all sorts of diagnostic procedures: x-rays, magnetic resonance imaging, drugs, other treatments. You need to know what's being done to your body. Lawyers need to understand an expanding arsenal of forensic methods. Think about it, genetic fingerprinting, spectroscopic analysis, drug testing, all these different kinds of techniques that you need to know if you're a lawyer. You can see in this recent newspaper advertisement, they are actually advertising now routinely that you can have genetic testing done on family members, paternity testing and so forth. Of course, if you're going to be a banker or an investor, you have to understand the nature of modern high-tech companies. There are so many areas that are really yielding tremendous profits. Think about genetic research, drug development, semiconductors, superconductors—the list goes on and on. Once again, if you don't understand those basic ideas that are part of every newspaper section, you're going to have a terrible time getting ahead in business. Legislators and voters, for that matter, need to be informed on issues about the environment and medicine, law and science funding, which is really important to me and my colleagues. That's the second reason.

The third reason to promote scientific literacy is your children. If you're scientifically literate, you can provide your children a tremendous foundation as they learn about science. Parents and teachers who are scientifically literate can reinforce children's experience, reinforce their curiosity, by looking at the world together. What happens when your child asks you a question about science? There's no better answer than "I'm not sure; let's find out the answer together." Scientific literacy provides the tools to learn so much more.

The last reason, but certainly not the least, the fourth reason to achieve scientific literacy, is that it allows you to share the joy of what I believe is humanity's greatest ongoing adventure of discovery and exploration. Every single day, scientists are discovering and sharing new things that no human being ever knew before. Through scientific literacy and through the range of subjects presented in this lecture series, you can share in that adventure. Everyone can achieve scientific literacy. Everyone can share the joy of science.

The great principles of science can be presented without relying on complex vocabulary. You don't need mathematical abstraction. We all live in a physical world of matter and of energy. Daily life has provided everyone with a wealth of experience in which to build understanding, yet in spite of this fact, several contrasting definitions of scientific literacy have confused and hindered science education reform. When I mention *scientific literacy*, some people immediately reply, "I can't even program my VCR." I argue that scientific literacy is different from *technological literacy*. That's the important, but very distinct, ability to use modern appliances, especially computers and other kinds of electronic devices.

Another common, but misguided, definition of scientific literacy comes from scientists themselves. The scientific literacy movement is not, as some scientists would advocate, just an effort to persuade more students to become science majors. Scientists will point to the pipeline problem. They say that science courses become more advanced and more students drop out, until at the end, only about 1 percent of all American citizens become scientists. The scientific literacy movement is not about increasing that number to 2 percent. On the contrary, I think educators should focus their efforts on achieving scientific literacy for all nonmajors. You don't have to be a scientist to appreciate the great discoveries and to find a joy in science.

Here's a copy of the National Science Education Standards. This document represents a strong consensus of thousands of scientists and educators regarding the content and the presentation necessary for all citizens to achieve scientific literacy. These standards, which Jim Trefil and I helped to write, provide a sound building code for science curriculum. The principle here is that scientific literacy has to be based on a few overarching principles rather than extensive vocabulary and lots of factual information. These principles come from all the branches of science, from physics and chemistry, from earth science, astronomy, and biology. Scientific literacy is also based on understanding something about science as a human endeavor. That means its history, its philosophy, its social context.

We also emphasize the scientific method, and that's a process of inquiry by which scientists ask and answer questions about the natural world. Here's an idealized version of the scientific method. You'll see here that it's a cycle. It's a cycle of observation, going from identifying patterns, to the hypothesis, to predictions. Predictions lead to more observations and data. And we're going to examine this process in a lot more detail in the second lecture.

For those of you who are teachers or parents, by the way, I've provided a concordance to the National Science Education Standards in your booklets. Every major content standard for grades K to 12 is covered in these lectures. The objectives of this lecture, this series on the *Joy of Science*, are to introduce the great principles of science and to explore how scientists discovered these sweeping ideas and to show how the great principles come into play in our lives. They're often surprising in very wonderful ways.

In these lectures, I'm going to attempt to integrate all the different sciences, and that's very unlike the traditional separation of physics and chemistry, of biology, of earth science into separate departments and into separate courses. I'm told that this division of science is a fairly recent phenomenon. It was, in large measure, the result of the European university system in the 19th century. Back then, each university department was allocated a single professor, and then there were lots of lecturers and assistants in each department. For a time, that worked out fine, because there were single departments of natural philosophy, or natural history. But as more and more people became distinguished scientists, the academic ranks swelled, and you had to have more professorships to keep everybody happy, so that meant more departments. This resulted in a rather arbitrary fragmentation of science. This got so bad that when I was in Cambridge University in the 1970s, believe it or not, there were three different earth science departments and there were seven chemistry departments, physically separated from each other for exactly this reason.

I think typical newspaper stories demonstrate this integration of the sciences in our daily lives. I showed you earlier a story on radioactive waste disposal. That's a great example, because think about radioactive waste. It's a key problem of the modern world, and you've probably read about these recent debates about whether to bury thousands of tons of nuclear waste at Yucca Mountain in Nevada. What is that problem? Is it physics? Of course, physics is important, because you can't understand radioactivity without a background in physics. Is it chemistry? Absolutely, because nuclear waste is chemicals, and chemicals react with the environment, so that's something we have to understand. Is radioactive waste disposal a geological problem? Absolutely. You have to bury these things for thousands of years in some stable environment, and that means rocks someplace. Of course, it's also a biological problem, because the only reason we're concerned about it is that radioactive waste affects people and other living organisms in adverse ways. That means that for a nuclear waste problem, if you got a degree in physics or in chemistry or in earth science or in biology, you may be able to chip away at one little piece of it, but you'll never be able to see the integrated whole of the problem. That's why integrated science is so important in our modern world. Virtually any problem in environment and health and resources is going to be integrated in that way.

For the final third of this lecture, I want to examine the social framework of the scientific enterprise. Science is a human endeavor and has its own distinctive organizational structures, just like any social system. First, let's look at the general enterprise of science in the United States, the whole picture. There are about 350,000 people who call themselves scientists that work in the United States. They work primarily in government, in industry, and in academia. Most of these scientists have a Ph.D., as well as several years of postdoctoral training, so it takes a lot work—just like a doctor—to become a scientist.

Scientific salaries are paid by universities, they're paid by corporations, they're paid by nonprofit organizations, and of course, they're paid by the government. A very significant fraction, in fact, does come from the federal government. In 2001, for example, the government allocated about $90 billion for research. That's our tax dollars at work. That's almost equally divided between defense-related and nondefense projects. The biggest spenders are the National Institutes of Health in Bethesda, Maryland (they take about $20 billion a year); NASA (about $14 billion a year); and the military—roughly half of the budget, $45 billion, goes to military research.

Science is a human endeavor, and humans invariably form groups. Scientists are no different, so let's look at a few of the very largest and most influential scientific organizations. There are a couple of organizations that cover all the different scientific disciplines, and the most influential of those is the Washington-based American Association for the Advancement of Science. It has about 130,000 members and is the major U.S. society that represents all the different branches of science. By the way, it's almost universally known as AAAS, and you may see newspaper reports about the

AAAS lobbying for science education. It also publishes *Science*, which is one of the most widely read magazines. *Science* comes out every week and has lots of news and features, as well as original science articles—it's very influential to get an article in *Science*. It has a rival magazine. This is a commercial venture called *Nature*, published in Britain. *Nature* also is a weekly journal with lots of science news and lots of original articles; it is very prestigious to get an article in a paper like this.

Another important organization is the National Academy of Sciences in Washington. It has a membership of about 2,000 scientists, so being elected to the academy is a very high honor. The academy still advises Congress on issues of science and technology. It publishes various reports, and they're summarized in something called *Issues in Science and Technology* that comes out monthly.

A third big organization that represents all scientists is Sigma Xi, a national science honorary. It's based in North Carolina. They support scientific research, and they have a monthly paper called *American Scientist* with general articles.

Let's look at the different specialties, because each specialty has its own set of societies and publications. This rigid division of science into separate departments may have hindered scientific literacy of nonscientists, but there is a rationale to the structure, and what results is a kind of hierarchy of scientific disciplines, of which physics is placed first. The laws of physics have to apply to all systems, large or small, living or nonliving, so physics can be thought of as this hierarchy atop of the hierarchy. Physics really can be defined as the study of matter in motion or, more abstractly, how particles interact. In understanding the great principles of chemistry, geology, and biology, therefore, you have to begin with an understanding of physics.

Physicists themselves divide their field into classical physics and modern physics. We have classical physics, which deals with everyday phenomena about matter and energy, forces in motion, things that we experience and great principles associated with these, which as we will see in the first few lectures, include Newton's laws of motion, law of gravity, laws of energy, the laws describing electricity and magnetism, and so forth. These are the things that we experience every day, and we find that life is a terrific classical physics laboratory.

Then there's modern physics, which deals with realms way beyond our daily experience: very small objects, very fast speeds, new physical laws that require quantum mechanics, nuclear physics, particle physics, relativity. These all become part of modern physics. Most of the 50,000 physicists or so in the U.S. belong to the American Physical Society. That's a big group that publishes *Physics Today*, a monthly publication. That group also includes geophysicists and astrophysicists, and other fields, so physics is a very diverse field in that sense.

Then we come to chemistry. Chemistry is the study of atomic interactions, especially the study of chemical reactions in the formation of new materials. Of course, modern chemistry is a very old field and is rooted in alchemy from the Middle Ages, but today, chemists follow the tradition in a different way. They mix chemicals together—they have this empirical sense to their science—but it's a very pragmatic science as well. You mix chemicals together and try to find something new; sometimes inspired guesswork, sometimes theory leads to these very exciting new materials. As a result, chemical research and chemical engineering are very closely linked. It's often possible to scale up an experiment that you make in a test tube, and within a few months, you make a huge plant that does the exact same thing. This link between chemistry as a science and chemical engineering is very close, and that results in the American Chemical Society, the largest of the societies having both scientists and engineers working together. Most chemical engineers belong to this society. It's one of the largest societies in the world, with 150,000 members. Those members receive a weekly magazine called *Chemical & Engineering News*. It's kind of interesting to see engineering and chemistry linked together in this context.

Then, we come to earth science, my own field, my favorite field. Earth science is the study of the origin, the present state, the dynamic processes of our planet, and sometimes, we add in other planets as well, because they behave sort of the same way. Unlike physics and chemistry, earth science deals with a very large, complex system that has a past, so it's a historical science. Earth science as a field is rather young. It grew out of the practical experience of prospecting and mining. When you think about it, around you, virtually every material which we use is made from earth materials, atoms that came from the Earth. Geology is a subdivision of earth science that is the study of rocks and soils and their distribution. Most geologists belong to the Geological Society of America, based in Boulder, Colorado. They publish a journal called *Geology* and the *Geological Society of America Bulletin*. These are monthly publications that are received by about 20,000 members of that society. There is also the American Association of Petroleum Geologists—that's based in Tulsa, Oklahoma, where there's lots of oil—and again, they have about 20,000 members.

Then, we come to geophysics. Geophysics is the study of Earth's dynamic processes, such as tides and hurricanes and earthquakes. The main society there is the American Geophysical Union, with 25,000 members. They publish a newspaper every week called *EOS*, and that has all sorts of news about geophysics. By the way, the government supports geophysics in a big way in various sorts of geological studies through the U.S. Geological Survey and the National Oceanic and Atmospheric Administration.

Then, we come to biology. That's the study of living systems, which are by far the most complex objects we know. Because of its importance to health and medicine, a lot more money goes into biology, and it's the largest of all the sciences. But biology is an extremely diverse discipline, because there are so many ways to study living things. I mean, think about how you might study an ant. How would you study an ant? You could study an individual ant, or you could use a microscope and look at the parts of the ant—the legs or the organs—then a higher power microscope to look at cells or even the molecules and the genes of the ant. Those are all different ways, but you could go the other way. You could look at the social structure of ants in their ant colonies or even the ecology, how ants fit into a larger picture. Biologists do a lot of different things.

There are many botanists, zoologists, microbiologists that are concerned just with naming new species. Other biologists focus on more dynamic processes, like evolution and genetics. One consequence of this great diversity is that there's no one overarching society for biologists. There are lots of specialized societies, and there are hundreds upon hundreds of professional journals in the field. The government's big support for biology comes from the National Institutes of Health. Remember, that's in Bethesda. Twenty billion dollars a year, your tax dollars. That's one of the largest single line items in the annual U.S. science budget.

Let me summarize. We've seen, in this first lecture, that science is a way of knowing about the natural world. It relies on reproducible observations and on experiments, and it differs from other ways of knowing, like religion, art, philosophy, and social sciences. Scientific literacy, which is the content of this course, encompasses those facts, the concepts, history, methodology that you need to understand scientific issues and to follow this great ongoing adventure. Everyone can become scientifically literate. Everyone can experience the joy of science. Finally, we've seen that science is conveniently divided into the disciplines—physics, chemistry, geology, and biology—yet in spite of this formal academic separation, the natural world is integrated; it knows no such boundaries. Every scientist, whatever his or her specialty, uses exactly the same methodology, called the *scientific method*, and that is the subject of the next lecture.

Lecture Two
The Scientific Method

Principle: *"The idealized scientific method is a cyclic process of inquiry, based on observations, synthesis, hypothesis, and predictions that lead to more observations."*

Scope:

Science, first and foremost, is a search for answers, but every answer must begin with a well-conceived question. Scientific questions are richly varied in scope and content. Most scientific questions can be divided into one of four broad categories: (1) existence questions ask what's out there; (2) origin questions explore how natural objects and phenomena came to be; (3) process questions examine how nature works; and (4) applied questions look for ways to manipulate the physical world to our advantage.

The scientific method provides a systematic framework for answering such questions. Four steps in the idealized method include: (1) collection of data through observations and experiments; (2) recognition of patterns in those data; (3) formulation of hypotheses and theories to systematize those patterns; and (4) predictions that can be tested by more observations and experiments. In the real world, the scientific method rarely provides an exact blueprint for research, and scientific progress is often made by more haphazard and serendipitous routes.

Outline

I. Scientific research is an outgrowth of human curiosity. Scientific progress depends as much on well-formulated questions, as on a catalogue of well-established answers.

  A. Many important questions are beyond the realm of science. Science addresses only those questions that can be answered by reproducible observations, controlled experiments, and theory guided by mathematical logic.

  B. Scientific questions are richly varied in scope and content. Most scientific questions fall into one of four broad categories.

1. Existence questions ask what objects and phenomena occur in the natural world.
2. Origin questions explore how natural objects and phenomena came to be.
3. Process questions ask how nature works. These questions are often closely linked to inquiries about origins.
4. Applied questions look for ways to manipulate the physical world to our advantage, whether curing disease, devising new materials, or modifying the environment.

C. Answers to old questions often lead to new questions.
D. Scientific questions are often interconnected in surprising ways. As scientists explore the most sweeping unanswered questions they often discover links between what seem at first to be unrelated topics.
  1. Plate tectonic models of the Earth's dynamic interior bear directly on our understanding of life's origin and evolution.
  2. Studies of ancient mass extinctions provide models for understanding the importance of the global environment.
  3. Scientific questions are also often linked by their philosophical approach. Reductionism is based on the assumption that systems can be understood by examining the behavior of fundamental building blocks. At the opposite extreme, collective systems display properties completely unlike those of their smaller components.
E. Some questions are not now scientific, but may be some day as our understanding grows (e.g., the nature of human consciousness).
F. Several factors prevent us from obtaining complete answers to many scientific questions.
  1. Experimental error: all measurements, no matter how accurate, contain some error.
  2. The uncertainty principle: at the subatomic scale, every measurement alters the object being measured (see Lecture Nineteen for a more detailed discussion).
  3. Chaos: many natural systems are chaotic, and thus are inherently unpredictable.

4. The speed of light: limitations are placed on us by space and time.

G. In spite of these and other inherent limitations on inquiry into the natural world, the methods of science provide the most effective and powerful tool we have to understand and modify our physical world.

II. The scientific method is a complex, variable, human process, which differs in detail from scientist to scientist, and from discovery to discovery. The method can be idealized as a cycle of observation, synthesis, hypothesis, and prediction.

   A. The first step in most scientific studies is the collection of data, including observations, measurements, and experiments.
   B. The second step is the recognition of patterns—the search for symmetries. Most scientists have a deeply held belief that there are regularities and patterns in the physical universe.
      1. Sometimes this step involves recognizing similarities among seemingly different phenomena, such as different forms of electricity.
      2. Sometimes this step is a mathematical synthesis, fitting disparate data into one type of equation, such as Kepler's discovery of elliptical planetary orbits.
   C. Once a pattern is found, the scientist will propose a possible explanation in the form of a hypothesis.
   D. A scientific hypothesis, theory, or law must lead to unambiguous and testable predictions, requiring a new round of observations. Consequently, a scientific theory can always be disproved by an unfulfilled prediction, but it can never be completely proved.
   E. At the center of this idealized cycle there is always a paradigm—a prevailing system of expectations about the natural world.

III. The scientific method is rarely followed as an exact cycle. Human imagination, intuition, and chance are vital elements of the process.

   A. The example of Dmitri Mendeleev and the periodic table of elements exemplifies the scientific method.

B. Often an anomaly leads to new insights.
   1. When anomalies are found that violate well-tested theories and laws, it usually means that the old theory or law is a valid special case of a more general law.
   2. An everyday example is provided by the "hypothesis" that all objects fall under the force of gravity. The anomaly of a helium-filled balloon leads to deeper understanding.
C. The scientific method is an elegant process for learning about the natural world, but it is neither intuitive nor obvious.

Essential Reading:

Hazen and Singer, *Why Aren't Black Holes Black,* Preface and Prologue.

Trefil and Hazen, *The Sciences: An Integrated Approach*, Chapter 1.

Supplemental Reading:

Barrow, *Impossibility*, Chapter 1.

Kuhn, *Structure of Scientific Revolutions*.

Questions to Consider:

1. Identify an unanswered question that is important to you. Does it qualify as a scientific question? If not, is the answer informed by science?

2. How might you apply the scientific method to questions that arise in your daily life, for example as a consumer?

Lecture Two—Transcript
The Scientific Method

All scientific research is an outgrowth of our insatiable desire to know things. We all want to know things about the natural world, and as we saw in the last lecture, science is the best way to discover how the physical universe works. This process of discovery employs the scientific method, which may be idealized as a cyclic process of inquiry based on observations, synthesis, hypothesis, and predictions that lead to more observations. Too often science educators focus on the most firmly established scientific answers, and in the process, this dynamic process, the creative process of asking questions is sometimes shortchanged.

In the last lecture, I told you about the National Science Education Standards, which call for science education reform centered on the idea of science as a process of inquiry. This lecture, the second in our series, therefore focuses on questions.

I have two objectives in this lecture. First, I want to consider the surprising nature of scientific questions themselves. Then I'm going to tell you about the scientific method by which scientists attempt to answer questions about the natural world. To begin with, I want to tell you about six different aspects of scientific questions, and the first of these is that many important questions are absolutely beyond the realm of science. Science cannot tell you the meaning of life. It can't tell you who to marry. It can't tell you whether to have children. It can't tell you if God exists. You see, science addresses only those questions that can be answered by reproducible observation, by controlled experiments, by theory guided by mathematical logic. Economist and philosopher Kenneth Bolding has said, "Science is the art of substituting unimportant questions, which can be answered, for important questions which cannot." Well, Bolding may have been speaking only half in jest, but it's certainly true that science can't answer many of those important questions.

Nevertheless, science informs many of the important non-scientific questions in our lives, including choices about health, about work, about relationships. In your decision to have children, you may want to know something about your family histories of genetic diseases, for example. Even if you want to know about the meaning of life and the existence of God, you need to understand the nature of the physical universe in which we exist, and that's something science can do for you.

Of course, there are also many questions that are scientific, and that leads to my second point, the second point about scientific questions. Scientific questions are richly varied in scope and in content. Scientific inquiry is as diverse as the natural world itself, but most scientific questions fall into four broad categories, and I want to tell you about these four categories of scientific questions. We have existence questions, origin questions, process questions, and finally applied questions. The existence questions ask: what objects and what phenomenon occur in the natural world. This has been a passion of scientists for hundreds of years. Scientific explorers of past centuries reveled in voyaging to exotic lands in pursuit of animals and plants, of mineral specimens and fossils. Charles Darwin cut his scientific eyeteeth as a naturalist in the voyage of the *Beagle*. Chemists isolated element after element. Physicians dissected diseased corpses. You have astronomers cataloging stars and physicists that scrutinize the unusual phenomenon associated with electricity and magnetism. We have continuing research of this kind today. Many scientists are now cataloging galaxies. They're sequencing genomes. They're isolating different kinds of viruses. They're finding new planets all the time, unearthing new dinosaurs, and eavesdropping on extraterrestrial life.

The second kind of questions are origin questions. These explore how natural objects and phenomenon came to be. Questions about the origin of the universe, the origin of the Earth, and the origin of life itself are among the most fascinating mysteries in science. They have to be answered with a plausible sequence of historical events. They have to be deduced from contemporary evidence, and they have to be constrained by natural laws. These are wonderful questions to sink your teeth into.

And then we have process questions, questions that ask how nature works. These questions are often closely linked to inquiries about origins. They explore how stars evolve, how rocks erode, for example. You can ask how cancer develops, how atoms interact with each other, how fungi reproduce. These are all kinds of process questions. Descriptions of the dynamic interplay and evolution of natural systems helps us understand the past and the present, and also may help us to predict what's going to happen in the future of these natural systems.

And then we come to the fourth kind of question, applied questions. These look for ways to manipulate the physical universe to our advantage—curing disease, devising new materials, modifying the environment are all aspects of applied scientific questions. Such questions are often quite specific; they're often rooted in technology. The search for safer cars, for example, for cleaner energy sources, cheaper fuels, longer lives. These are all applied questions that many scientists are tackling today.

These pursuits, of course, are also closely tied to other kinds of scientific questions. You can't manipulate nature without first understanding it. Curing infectious disease, for example, that requires the knowledge of infectious organisms: what are out there, how they behave, why they behave the way the do. And, of course, constructing skyscrapers involves understanding the properties of materials, a detailed knowledge of how these materials behave under various stresses.

Now to a third aspect of scientific questions, and that is that old questions, when they are answered, often lead to a whole new set of questions. In 1843, the art critic John Ruskin lamented that to know anything well involves a profound sense of ignorance. This sentiment strikes a very responsive chord in most scientists today. The more knowledge grows, you see, the more we realize how much we don't know. There have been many unanticipated discoveries, like x-rays, tectonic plates—which I'll tell you about—the genetic code, buckyballs. All of these discoveries opened up vast new areas of research. Rather than answering a question, you really just opened the door on whole new regimes. The answer to existence questions invariably leads to questions about process and origin. For example, if we were to discover extraterrestrial life, it would not only tell us that those life forms existed, but would open up huge areas of research about exobiology, which we can't even conceive of right now.

My fourth point about scientific questions is that these questions are often interconnected in surprising ways. If scientists explore the most sweeping unanswered questions, they often discover links between what had first seemed to be very unrelated topics. Big questions really then tend to blur the traditional boundaries of science departments that unify our view of the natural world.

Let me give you some examples—plate tectonics. Plate tectonics was a revolution in the earth science, which showed how the dynamic interior of the Earth bears directly on our understanding of what's happening at the surface. But it goes beyond that. Plate tectonics also provided profound information about life's origin and its evolution. Then we have studies of ancient mass extinctions. These studies provide models for understanding the importance of today's global environment and may help us answer many of the questions that we have about what humans are now doing to the environment—even though we're looking at environments on the globe millions or hundreds of millions of years ago.

Scientific questions are also often linked by their philosophical approach. For example, we have reductionism, which provides one approach to answering questions. It's based on an assumption that physical systems can be understood by looking at the smallest building blocks, the smallest units. You can understand materials by looking at atoms. You can understand people by looking at their genes. Particle physicists search for the "theory of everything," which I'll talk about in a later lecture. Here's another example of this, looking for the finest structure of matter, even below the scale of the atoms, below the scale of the atomic nucleus, and from that trying to understand the entire physical universe. That's an example of reductionism.

At the extreme opposite, you have ecosystems, you have the brain, the universe as a whole, and these are vast systems that have to be studied in their collective whole as unity. It turns out that many of these large complex systems display behavior that you could never predict just on the basis of the individual components. You see what are called emergent properties. From atoms you get planets. From planets you get life. From life there emerges consciousness, and at each stage you never predict that emergent property from the previous set of starting materials.

A fifth aspect of scientific questions is that some questions are not now scientific, but someday they may become so. Let me give you an example from history. In 1900, before Edwin Hubble's discovery of the galaxies, questions about the origin of the universe were not in the realm of science, they were pure philosophical speculation because there's no observation you could make about the universe that would tell you anything about its origin. Nobody had thought of a kind of observation that would help. But when Hubble discovered distant galaxies, and discovered that they were moving away from us, it opened this whole new realm of questioning about the origin of the universe, and we'll see that in a later lecture.

Today, the study of human consciousness is probably beyond science as well. We don't even have a good operational definition of consciousness. We don't know how to quantify it. We don't know how to define it. We certainly don't know how to study consciousness as any physical entity. But that situation may change. After all, many people are studying the brain, studying the behavior of neural systems, and so someday as we understand more and more, the question of what is consciousness may fall into the scientific realm.

Now to my sixth and last point about scientific questions, and that is that most questions can't be answered completely. This is a great frustration to science. There are at least four factors that prevent us from obtaining complete answers to many of our scientific questions. Let me tell you about these four factors.

The first is a common one that you've all heard of, and that's experimental error. All measurements, no matter how accurate, no matter how carefully done, contain some error. Think, for example, about measuring the thermal expansion of a metal bar. If you heat up a piece of metal, a bar for example, that is about three feet long, it is going to get a little bit larger. And thermal expansion is an exact measurement of how much longer the bar gets with temperature. You can measure thermal expansion. You can measure the length of the bar and you can measure temperature, and so you can have a plot, temperature versus length, and that tells you thermal expansion. But no matter how careful you do those measurements, there's always going to be some error in the measurement of temperature, no matter how well you design your thermometer, there's going to be some uncertainty. There is also going to be some uncertainty in the length. Maybe you can measure that length to a thousandth of an inch, but can you measure it to a millionth of an inch? If you can measure it to a millionth of an inch, can you really measure it to a billionth or a trillionth of an inch, or more? There's always going to be some degree of error. There's always room for improvement in scientific experiments, and so the scientific process is never ending, and some questions can never be completely answered.

Now, science is a never-ending process because new techniques for making these measurements are constantly being invented, so we can always improve the precision and the accuracy with which we make our observations and measurements. We have bigger telescopes. We have faster electronics. We have more sensitive analytical methods, for example. This regularly improves our ability to make measurements, but we're never going to reach the end of that road. There's always room for improvement.

A second reason why scientific questions can't be answered is called the uncertainty principle. And this applies to the subatomic scale. At that scale, every measurement that you can conceive of making alters the thing that's being measured. This is one of the most unsettling discoveries of quantum mechanics, because it turns out at nature's smallest scale, the scale of the electron, the scale of the atom, everything that you can use to measure atoms and electrons has comparable energy to the atoms and electrons you're trying to measure. It's sort of like trying to determine if there's a bowling ball in a room which is dark, and the only way you can try to find the bowling ball is by rolling other bowling balls into the room and listening for a clunk. But as soon as you hear that clunk, the bowling ball has moved its position, and you don't know where it is anymore, and that's sort of what happens at the subatomic scale. The uncertainty principle is another reason why anything you measure is going to change so we can't know everything there is to know about the physical universe.

Chaos is a third problem. Many natural systems are chaotic, and thus inherently unpredictable. We believe that there is an orderly universe. We believe in natural laws in which a given set of initial conditions leads to a predictable outcome. But the trouble is, these chaotic systems are such that the slightest change in initial conditions some place down the road can affect the systems and their behavior drastically. One of the ways this is described metaphorically is called the butterfly effect. The idea is that a butterfly, flapping its wings in Asia can change the weather pattern over North America several months later. So even if an entire physical system obeys all the physical laws, and even if we can codify all those physical laws, we can never measure the initial state of a system with such precision and accuracy that we can predict how it's going to become if the system is chaotic. So that's another inherent limitation on knowing how the universe is going to behave.

And a fourth, and I guess a rather prosaic limitation on what we can know, is the speed of light. To the best of our knowledge the fastest anything can travel, any physical object can travel is just under about 186,000 miles per second, and that means there are some places in the universe we're never going to be able to go. There are distant objects for which we cannot possibly know what's happening to them today. We can only see them because they are far away as they were in the past. It also means that we can never travel backwards in time, at least with present technologies and present understanding of the physical universe. And these are other limitations on what we can know. But I have to say, in spite of these and other inherent limitations into the natural world, the methods of science absolutely, positively provide the most effective, the most powerful tool that we have for understanding and for modifying the physical world.

I now want to change gears. I want to tell you about the scientific method. This is a systematic process that scientists have developed to discover principles of the natural world. Now don't get me wrong, the scientific method is a complex, it's a variable process, it's a human process, it differs in detail from scientist to scientist, from discovery to discovery. It's very hard to give you a simple overview of the scientific method, and one of the things I want you to get out of this lecture series is how richly varied the discovery process can be. But having said that, it is possible to idealize the process of scientific method into a simple four-step cycle, and these steps are observation, synthesis, hypothesis, and prediction. The predictions then always lead to new observations so the cycle goes on and on.

The first step in most scientific studies is the collection of data. This includes observations, it can include measurements, it can be experiments of various sorts, and these data have to be tabulated. They have to be expanded into monographic studies of all the different kinds of objects or all the different kinds of phenomena that are out there. They can be collected specifically to test a prediction, but primarily you're just going out there and seeing what's out there. This is related to those existence questions I was telling you about. Many, many of the scientific text that you've seen relate to this sort of effort. For example, you might want to describe all of the different bird species in a particular area, and so eventually you bring these bird species together, and you'd have books that show lots and lots of species. You've seen these, they're available to amateur bird watchers, but this is an essential part of the scientific effort to catalogue all the different kinds of species.

One of my very favorite books of this sort, because my wife and I love to collect fossils, is a book that described all the known trilobite species. And this, again, is a bible for people who are involved in the paleontological study of trilobites, page after page after page of all the known varieties of trilobites. These are ancient fossil animals that lived more than 300 million years ago. So that's the first step, collecting data.

Now, sometimes these data are going to be tabulated in huge volumes of numbers. For example, I told you about thermal expansion before. You can have table upon table of thermal expansion coefficients for all different kinds of materials. That information is vital, both for engineers who are trying to build structures like buildings and bridges, but also for scientists trying to understand the properties of materials. And these tabulations, these volumes of numbers are an essential part of science. They are the first step in that scientific method.

The second step is the recognition of patterns—the search for what I call symmetries. Most scientists have a deeply held belief that there are regularities, that there are patterns in the universe, and we can find those patterns. The recognition of a pattern in nature from its vast amounts of data, that's the art of science. Sometimes this step involves recognizing similarities among seemingly different phenomena, such as different forms of electricity. Indeed, one of the great discoveries in the 19th century was that lightning and static electricity and the kind of electricity you produce by electric eels and the kind of electricity produced in batteries are all one and the same phenomenon. Recognizing these very different physical phenomena as being one sort of thing, that's a process of synthesis.

Sometimes the step is mathematical in nature. Finding all sorts of disparate data and realizing that they all fit one simple equation. And I'll talk to you about Johannes Kepler and how you recognize that the orbits of planets fit this kind of mathematical relationship of form called an ellipse, and that was another great discovery.

Now the third process. Once a pattern is found, the scientist is going to propose a possible explanation in the form of a hypothesis. Often this step is just as straightforward as generalizing the synthesis. Kepler saw one planetary path was an ellipse, and he made a generalization that all planetary paths are ellipses. You see that two types of electricity are the same, and then you say that all types of electricity are the same, and so forth. Once a hypothesis is well tested, once its examined by many, many people and seems to fit all the observations, it may assume the status of a natural law, that is if it's mathematical in form, or perhaps a theory if it's more descriptive in nature, as in Darwin's theory of evolution by natural selection.

A key point about scientific theory is a hypothesis. Any scientific hypothesis or theory must lead to unambiguous predictions. They have to be testable predictions; they require a new round of observation. If you propose a theory and there's no way to test it, then it's beyond the range of science. Consequently, a scientific theory can always be disproved. If you make a prediction and that prediction is unfulfilled, then your theory is wrong, or at least the theory has to be modified. You can never completely prove a scientific theory, a scientific law, because there's always some measurement yet to be made that's a prediction of that theory. Science is, therefore, extremely vulnerable to attack. A single odd fact has the potential to undermine a long-standing theory. Each counterexample has to be laboriously scrutinized. You have to go into the laboratory and say, why does this example not seem to fit that theory. And this is why supporters of evolution often find themselves on the defensive when they're debating creationists. Creationists say evolution did not occur and they can show individual examples of rocks that may seem to violate the concept of evolution. For example, they might show you a rock with limestone and fossils that are supposedly hundreds of millions of years old, and yet they have human footprints in the same rock. Well, that would violate any idea of the theory of evolution of humans being much more recent than the hundreds of millions of years old rock. So each such anomaly then has to be analyzed and scrutinized. It's very difficult to argue against such evidence in a two-hour debate. You need years and years to analyze this sort of data.

There actually is one more part to this four-part cycle, and that's right at the center, that's the idealized cycle always has to have a paradigm in the middle. The paradigm is the prevailing system of expectation—the prevailing ideas about how the natural world works—because every observation, every prediction, every hypothesis is made in the context of our present understanding of the natural world. So underlying all of this human process of the scientific method, there always is an underlying paradigm. We wouldn't ask questions about life on other worlds, for example, if we thought that life on Earth is a unique phenomenon. If that was part of our built-in belief structure, you just wouldn't think about life on other worlds. And I'll bet you that there're many important and interesting questions that scientists could ask, that they're not asking right now just because it doesn't fit into our current paradigm. Paradigms, therefore, gradually shift and the question scientists ask also shift.

I should tell you right up front that the scientific method is rarely followed as an exact cycle. Human imagination, intuition, and chance are vital elements of this process, and we're going to explore some of those variables in the next lecture. But first I want to give you a historical example, which is often cited as the epitome of the scientific method at work, and that's Dmitri Mendeleev's discovery of the periodic table of the elements. Mendeleev had data on 63 different elements. That was his input; those were his observations. That was enough to begin to see patterns. He expected that he was going to find order in this collection because he believed you couldn't have 63 independent and separate fundamental units of matter. He thought there had to be some underlying order, and so he looked for it. That was his paradigm. He ordered the elements from left to right according to their weight. He ordered them in vertical columns when he saw elements that had very similar behavior. And so he built up this table, and everything seemed to fit. Furthermore, the table that resulted had a few gaps in it, places where it seemed like elements were missing. So he predicted that those elements would be found, and sure enough within a decade after he predicted the table, they actually found some of those elements that were missing. That is, in a nutshell, how this whole scientific process works.

Now, the Mendeleev story is compelling, but science doesn't always work quite in such a linear fashion. Often there are anomalies that lead to new insights. Historian of science, Evelyn Fox Keller has observed, and this is a quote, "when scientists set out to understand a new principle of order, one of the first things they do is look for events that disturb that order." Almost invariably it is in the exceptions that they discover the rule.

When anomalies are found that violate well-tested theories and laws, it usually means that the old theory or law is valid, but it's just a special taste of a more general law. Let me show you what I mean. An everyday example would be provided by the hypothesis that all objects fall under the force of gravity. I can show you this. I can take an object and drop it, and it falls. I can take another kind of object, drop it, and it falls. Another object and drop it. I could do this all day long, take objects and drop them and they fall. It's part of everyday experience, and so you develop a hypothesis that when objects are dropped, they fall. And then, in your process of studying the universe with a great deal of care and precision, you finally drop a helium-filled balloon, and it doesn't fall. What does this mean? How do you accommodate this into your theory? It doesn't mean that the idea when objects drop that they fall is wrong, it means that the falling objects are part of a more general theory. Indeed, it's Newton's theory of gravity, which explains both the objects that fall down, and the objects that fall up. And then if you look in more detail about how things fall, you find that Newton's laws are only a special case of Einstein's more general relativity in which objects that are moving very fast or are very massive behave differently than the kind of objects we have in our everyday world. So the classic laws of motion fail under conditions of extreme speeds, extreme masses, and so forth, and so you keep having to have more general laws that subsume the more familiar everyday kind of experience that we have.

Geneticist Barbara McClintock underscored this point in her Noble Prize winning work on corn genetics. Other workers at the time believed a simple relationship existed between individual genes, which were passed down from parents to offspring—genes that determine the characteristics of individual corn kernels. They sort of thought of the master gene—a gene controlled the trait. Well, McClintock thought differently. She saw the kernels in a different way. She said, "See that one kernel that was different in your corn, and you've got to make that kernel understandable. And when you see the one kernel that's different and understand that, then you find the truth about genetics." In the process, McClintock was able to find that the genes occasionally moved, they shift places, they jump, and this provides an additional vitally important mode of genetic variation that no one had seen before. So the scientific method is not a strict guidebook, but rather it's an idealized path to knowledge that often seems to fit better in retrospect, as you look back on history, than it does while the discoveries are actually being made.

Now let me summarize this lecture. First, I considered six aspects of scientific questions. First, many important questions are beyond the realm of science. Second, scientific questions are richly varied in scope and in content. Third, answers to old questions often lead to new questions. Fourth, scientific questions are often interconnected in surprising and remarkable ways. Fifth, some questions are not now scientific, but may be someday as our understanding grows. And sixth, most questions in science can't be answered completely.

Then I looked at the idealized four-step scientific method—which includes first, observation and experiments, that is, collecting data; second, recognition of patterns in that data; third, the development of a hypothesis or a theory; and fourth, the predictions which then lead to new observations. This scientific method is an elegant process for learning about the natural world. It's neither intuitive nor obvious, however. In the next lecture, I'm going to chronicle the 5,000-year development of the scientific method.

Lecture Three
The Ordered Universe

Principle: *"Many aspects of the physical universe are regular and predictable."*

Scope:

Scientists share the belief that our senses don't lie. Such an assumption was not obvious to ancient societies, who were subject to the devastating uncertainties of weather, disease, famine, and war. But primitive people did grow to depend on certain regularities in nature: the sun rises and sets, rocks fall when dropped, seasons flow in a regular progression. The Roman scholar Pliny the Elder (AD 23–79) represents a more advanced stage in the development of science. He catalogued thousands of "facts" that were known to him in his *Natural History*.

One of the earliest applications of the scientific method came from efforts to model the motions of the Sun, Moon, and planets. Ptolemy of Alexandria (c. AD 100–170) proposed an Earth-centered model that incorporated circular orbits modified by secondary circles, called epicycles. The Ptolemaic system was accepted for nearly 1400 years, until improved observations revealed discrepancies in the Ptolemaic predictions of planetary positions.

Outline

I. The idealized cycle of the scientific method— observations, synthesis, hypothesis, and predictions leading to more observations—seldom works. Science is a human endeavor characterized by luck and intuition, as well as mistakes and misperceptions.
   A. Honest errors in technique or execution are an inevitable part of the scientific process. The demand for reproducibility eventually eliminates these mistakes.
   B. Scientific fraud is rare because the scientific method insures that such cases will eventually be discovered and corrected.

II. To modern scientists, each observation and measurement represents an objective, verifiable truth; each measurement thus has meaning. But the importance of observation has not always been obvious or widely accepted.
   A. The Greek philosopher Aristotle (c. 384–322 BC), who was a student of Plato and tutor of Alexander the Great, recognized that the senses play an important role in understanding the natural world. But to him and to his followers, logic and reason, rather than our fallible senses, were the ultimate arbiters of truth.
   B. Many scholars prior to about 1650 accepted the writings of Aristotle and other ancient authorities in preference to observations.
   C. Galileo met resistance in the early 1600s when he first used a telescope to observe the heavens. After he discovered supposed "imperfections" in the heavens (features such as sunspots and craters on the Moon), some of his contemporaries refused to look through the telescope for themselves.

III. While many scholars have questioned the reliability of the senses, everyone relies to some extent on the predictability of the physical world.
   A. The simplest kind of verifiable observations are events that happen over and over again in our lives.
      1. Objects fall when dropped and adopt predictable curving paths when thrown.
      2. Most astronomical objects also have predictable paths.
   B. Some ancient societies displayed their belief in the predictability of the Sun and Moon by constructing large monuments in which the positions of massive stones were aligned with key events in the calendar.
      1. One of the earliest examples of human's faith in the predictability of nature is found at Newgrange, Ireland, the site of a 5000-year-old burial tomb of an ancient Irish leader.
      2. Stonehenge, the great monument on southern England's Salisbury Plain, served as an elaborate and sophisticated calendar.

- C. Science depends on maintaining meticulous records of observations. Such was the lifelong task of the encyclopedist, Gaius Plinus Secundus (AD 23–79), better known as Pliny the Elder.
  1. Pliny led a rich and varied life as a cavalry officer in Germany, a naval commander at Naples, a lawyer, and an administrator in Rome's Spanish colonies.
  2. He is best remembered for his encyclopedic *Natural History*, which is divided into 37 books, each of which tabulates facts known to him directly or through reliable sources.
  3. Pliny died while observing an eruption of Mt. Vesuvius in AD 79.

IV. To many ancient cultures, knowing the positions of the Moon and planets was of great importance both for astrological predictions and for navigation. The thousands of visible stars form a constantly rotating backdrop of constellations against which planetary motions can be plotted.
  - A. The position of a planet can be measured night-by-night by its position relative to the fixed stars. Planetary motions are complex. Mars, for example, appears to move forward through the Zodiac for much of the year, only to slow and backtrack every few months. Explaining this "retrograde" motion was the major challenge in devising models of the solar system.
  - B. One of the first descriptions of the heavens that led to testable predictions of planetary positions was the extremely successful model of Ptolemy of Alexandria (c. AD 100–170).
    1. Ptolemy, an Egyptian-born Greek, collected his own observations along with those of earlier Babylonian and Greek astronomers.
    2. Ptolemy argued that the Earth must be at the stationary center of the universe.
    3. The Ptolemaic model postulated perfect circular orbits on which smaller circular paths (epicycles) were superimposed. The use of circular orbits was dictated by a belief in the perfection of the heavens.

4. In spite of its complexity, the Ptolemaic system was reasonably successful in predicting planetary positions, and it was employed for more than 1400 years. It was not until the 1500s that new, more precise observations and application of the scientific method led to general acceptance of a competing model.

Essential Reading:

Trefil and Hazen, *The Sciences: An Integrated Approach*, Chapter 2.

Supplemental Reading:

Cohen, *Birth of a New Physics*, Chapters 1–3.

Hawkings, *Stonehenge Decoded*.

Questions to Consider:

1. Imagine yourself living in a primitive society. How might you attempt to convince other people that the universe is regular and predictable? What might be their counter arguments?
2. Given the observations that you can make yourself without the benefit of a telescope or space travel, is it more reasonable to conclude that the Sun orbits the Earth, or vice versa?

Lecture Three—Transcript
The Ordered Universe

Scientists study the natural world using the scientific method. But the scientific method didn't spring forth fully formed. It's not at all an intuitively obvious process. In fact, it emerged gradually as humans began to realize the great scientific principle that many aspects of the physical universe are regular and predicable. In this lecture, I want to tell you about the slow, but seemingly inexorable path towards our present view that the universe is governed by a few great overarching natural laws, but first I want to look at the scientific method in somewhat more detail.

Recall that the scientific method has four parts. First, you have observations, you have experiments, you collect data about the natural world, and you systematize this data in tables. And then second, you begin to see patterns. You begin to see systematic behavior in those data, and this leads to the third step, that's hypothesis, theory, making up the natural laws. But every hypothesis, every theory has to make predictions, and that's the fourth step in this scientific method, you make predictions, and then you can go out into the natural world and you make more observations. You conduct more experiments, and so the cycle goes on and on and on continuously.

Well, that's the idealized scientific method, but scientific discovery seldom follows this exact idealized path, so I want to look at the scientific method in a little bit more detail in some of the classic examples. Even the case of Dmitri Mendeleev and his discovery of the periodic table doesn't really follow that exact pattern. Let me tell you why. Mendeleev had 63 chemical elements. Those were his data. He saw patterns in those data. He saw systematic trends. So for example, rubidium, potassium, sodium—three elements that are all soft, silvery metals, they all react violently with water, they all combine with chlorine and fluorine in one-to-one ratios—these he grouped together. He said they're going to be a vertical column of elements because they're so similar. But, Mendeleev had to ignore equally obvious and compelling similarities in other elements. For example, think about iron and cobalt and nickel. These are all silvery metals. They're all quite hard. They all react in a one-to-one ratio with oxygen. They're all magnetic, and that's a striking property because very few chemical elements are magnetic, but all three of these are. And yet, Mendeleev put these three elements side by side in his table. He ignored their similarities, and did not put them in a vertical column. So he had to have some kind of intuitive hunches about how you put the table together. It was not just a straightforward process. It involves some creative leaps.

Another great example of an intuitive leap is that of Isaac Newton and his understanding of the motions in the heavens and on Earth. In the 1660s, Newton's contemporaries saw all the motions of planets and stars in the heavens as one sphere, as one domain, completely distinct from motions on the Earth, but Newton had another intuitive idea. He said, perhaps these motions are in some way related. Perhaps when you drop an object and gravity brings it down to Earth and then when you throw an object it adopts a curving path. But if you could throw it really hard, if you could take an object and throw it as hard as you possibly could, so that it actually went into orbit around the Earth, then it would adopt the same curving path that you see characteristic of the Moon, of the stars, of the planets. And so, Newton, unlike any of his contemporaries, saw that similarity. He saw the connection between motions on Earth and motions in the heavens. That was the great intuitive leap. No one else had ever seen the universe that way.

These stories of intuitive leaps often take on mythic proportions. You have the story of Einstein sitting on a trolley car in Switzerland receding from a clock tower and thinking about relativity and the story of Newton, of course, sitting in an apple orchard with the apple falling. Some people say it fell on his head, of course it didn't, but that's the story. Scientists claim to have these intuitive moments, these great insights in the shower. They have them in dreams, and it's always this critical nonlinear aspect of the scientific process that's involved. Something you really can't legislate by a strict scientific method.

Science is intensely creative. You know I'm often amused at people's reactions when they find out that I'm a musician. I play trumpet in a symphony orchestra. People say, "Oh, you must be so creative because you play a musical instrument." It's just absolutely not true. Music is a discipline. When I go into an orchestra, I have to play the notes that are on the page. I have to play when the conductor says. I have to play at the dynamic the conductor says, because a trumpet player gone awry who is too creative can really wreak havoc in a symphony orchestra. That's the discipline in my life, but science—I go into a laboratory, and think about it, I've got chemicals, I've got apparatus, I can do anything I want to. Nobody's telling me the next experiment to run. Nobody's telling me how to synthesize that data. And so science is intensely creative, because you have to constantly make it up as you go along, and the scientific method only gives you the loosest, vaguest sort of guide as to how you might do that.

There are other human aspects of science that add to its interest and complexity. One is that scientists often make mistakes. But, science is a self-correcting process. It's a collective process in which anytime you make a finding, someone else is going to come along and try to verify that finding. Honest errors in technique and execution are just part of the game. It's inevitable. The demand for reproducibility eventually corrects and eliminates these mistakes. There's a great example, provided by a recent report that came out of Tulane University in the chemistry department. They published a very widely cited report about the intensified environmental effects when you combine certain pollutants. Here is the situation. If you have one part in a million of pollutant A, it becomes dangerous. If you have one part in a million of pollutant B, it becomes dangerous. What the Tulane researchers claimed is if you put the two chemicals together in much lower concentration they were, because of their combined effects, much more dangerous than they would be individually. And this is a terrible problem, for example, for the Environmental Protection Agency, because EPA sets toxicity limits for various chemicals. But imagine if you not only have to define the toxicity of thousands of chemicals, but then every possible combination of those chemicals. You get millions or billions of different combinations, and it's just an impossible task to identify those risks. So this was big news. They published the article, but then they tried to reproduce it and do more experiments, and the Tulane researchers themselves were unable to do it. So after all this publicity, they had to retract some of their findings. That's what scientists should do, but nevertheless, certain science journalists accused them of fraud. This was not fraud. This was just the normal process of science. Every complex experiment has many variables—many things that are difficult to control—and sometimes procedural errors happen. It doesn't constitute scientific fraud. It's just a part of the process.

Later in this lecture, I'm going to tell you about the famous case of cold fusion, in which two Utah researchers claimed that they had a new source of energy. Well, they were wrong, and they were justly embarrassed, but once again, this wasn't fraud, this was just the kind of error that humans make, and the kind of error that humans become very embarrassed once they do make.

Having said that, I really should address this question of scientific fraud, because once in awhile there are cases. It's rare, but it does happen. And once again, the scientific method is a correcting process. So I want to tell you the story about an American paleontologist named Albert Carl Koch, who lived in St. Louis in the 1830s and 1840s. Here's a good example of a fraudulent bit of science. In this period, in the 1830s and 1840s, there were fossil bones in rich abundance in various remote parts of North America. Albert Carl Koch made it his passion to go out with an ox cart and just ride along the countryside, find one of these deposits, and fill up his ox cart with fossil bones. For example, he had fossil bones of mastodons, and fossil bones of whales. He'd pick up bones. Some of them six or eight inches across, some of them vertebrae up to a foot or more across. He'd bring them back to his St Louis museum where he charged admission for the various exhibits that he put together. Let me read you a quote by Koch, who describes his collecting process. "The place where those relics were deposited is one of the most romantic situations I have seen in Missouri. The bones in general had been broken. Those that have been thus far been disinterred consist of, at first, the head of an undescribed animal from which it appeared that it exceeded the elephant in size from four to six times." This is in one of Koch's scientific publications. What Koch had done was that he had gotten as many bones as he possibly could, many vertebrae, many ribs, and then he'd stack them together end to end. He'd make long chains of them, as many vertebrae as he could from all different species. He'd stick them together. He'd make the most outrageous concoction he could. Indeed, organisms that were much larger than the elephant today. He would make fossil sea serpents stringing hundreds of vertebrae together making individual fossil skeletons more than 100 feet long. And he could charge admission for these—$0.25, $0.50, that was a lot of money in those days. And people would flock to see these remarkable, fanciful reconstructions.

Koch abused the scientific method. His data were the bones. His hypothesis was these reconstructions that he knew were false because he had taken bones from different places, strung them together in ways that were physically impossible. And yet, when scientists saw these, they denounced the reconstructions, but they purchased his fossils from him. They took them back to museums, and they reassembled them in ways that were correct, they were anatomically reasonable. And you can go to museums today, to the British Museum, for example, and actually see some of the bones that Albert Carl Koch collected because science is a corrective process. The data were still good. They were assembled in a non-fraudulent way by the museum, and now we can benefit from Koch's collecting.

No matter how compelling a theory is, it has to stand up to repeated tests, to scrutiny of the scientific community. And here I want to come to a central point. Science is not a religion. Scientific truth is based on independently verifiable observations, on experiment, and on logic. But scientists do share a belief. They take it on faith that our senses do not lie. What we see, what we touch, what we smell, what we hear, these are truths, and if I can verify it and you can verify it independently, that has a kind of truth to it. You can make a measurement. I can repeat that observation, and our answers should match within reasonable experimental error. In the long run, fraud in science, therefore, just doesn't pay.

Let's take a closer look at this central belief, or assumption of science. To modern scientists, each observation and each measurement represents an objective, verifiable truth. Each measurement thus has meaning. But the importance of observations has not been at all obvious or widely accepted in earlier times. The Greek philosopher Aristotle, who lived from 384 to 322 B.C., is a good example of this. He was a student of Plato. He was a tutor of Alexander the Great. Certainly he recognized that the senses played a very important role in understanding the natural world. But to him and to his followers, logic and reason, rather than our fallible senses, were the ultimate arbiters of truth. It was the brain, not the hands that led to an understanding of the natural world. In the words of the British historian Neville Andrade, "Aristotelians considered that knowledge about nature could be won by pure thought, and that experiment was, in a way, a rather trivial and pedestrian proceeding."

Many scholars prior to about 1650 accepted the writings of Aristotle and other ancient authorities in preference to observations. In our textbook, *The Sciences*, Jim Trefil and I recount a story about a debate at Oxford College that occurred during this period. And this debate was about the number of teeth in a horse's mouth. One scholar quoted Aristotle. Another countered with a different opinion from the theologian St. Augustine. And so forth, back and forth the debate went. Eventually, so the story goes, a young monk at the back of the auditorium got up and noted, that since there was a horse outside, they could just go out, look in the horse's mouth, and count the number of teeth. At this point, records claim, the assembled scholars "fell upon him, smote him hip and thigh, and cast him from the company of educated men."

The great Italian scientist, Galileo Galilei met a similar kind of response in the early 1600s when he first used the telescope to look into the heavens. Galileo had heard about the telescope as a Dutch invention. He made his own in 1609 and focused it on various objects in the night sky. He discovered supposed imperfections in that night sky. He found moons orbiting Jupiter. He found sunspots. He saw what appeared to be a vague fuzzy mass around Saturn, what later turned out to be the rings of Saturn, and so forth and so on—craters on the Moon, mountains on the Moon, imperfections in what should be perfect spheres in a perfect heaven. Historian of science Giorgio de Santillana writes, "Galileo had thought the discoveries of the telescope would provide irrefutable proof to any man in good faith. But a few months were enough to undeceive him. Surgeon doctors steadfastly refused to look through the telescope. Some did look and professed to see nothing. Most of them, however, said that they had never gotten around to looking through it, but that they knew already that it would show nothing of philosophical value."

Let's shift gears now and look at the nature of predictability in the universe. While many scholars have questioned the reliability of the senses, everyone relies, on some extent or another, on the predictability of the physical world. The simplest kind of verifiable observations are events that happen over and over and over again—things that occur in our lives with regularity. Long before science, humans developed an absolute dependence on certain repeating patterns in nature. For example, the way objects fall, and the way they move when they adopt curving paths through the air is absolutely critical to a hunting society. So if you take an object and drop it, you have confidence that it is going to fall. If you take the object and drop it again, you say, "It fell again." There's a repeating pattern. Or when you take an object and throw it, it adopts a certain kind of curving path. If you take another object and throw it again, it adopts the same kind of curving path, and you rely on this, for example, if you are a spear thrower and you're hunting game, you rely on that ability to throw things in the same kind of arc again and again and again. This is vital to the survival of any people.

Most astronomical objects also have predictable paths. The Sun rises and sets each day, while its position in the sky and the lengths of the day change in a regular pattern from season to season. The Moon also changes position in a regular fashion, and it also passes through a predicable 29-day cycle of phases, from new moon to full moon, back to new moon, and so forth. You can measure these changes yourself, and you can do it in a very simple way. This is a wonderful thing to do with your children if you've never studied the phases and the shifting patterns of the Moon, it's fascinating to do this, and you can graph your results in a very simple way. Let me tell you how to do this. Pick a time of day. My favorite is 7:00 a.m. or 7:00 p.m. Since the Moon changes its position systematically, some weeks you're going to have to look in the morning sky, some weeks in the evening sky to see the Moon. And here's what you do. You pick that time. You have to have an accurate watch. Always do it at 7:00 on a particular day, if you choose that time, or 8:00 or 9:00 whenever it's convenient for you. Take a compass. Sight the direction to the Moon, and take a compass reading from north, a certain number degrees to the west of north or to the east of north, that's where the Moon lies in the sky. That's the first number you need. That's the number along the horizon. And the second thing you want to do is measure the height. And the easy way to do that is to take a protractor that's marked off in degrees. I like to tape a straw to the edge of the protractor; that gives you something to sight along. You then take the protractor, sight up to the angle of the Moon, and let a little weight hang down. Hold that weight, and measure the angle. Do this day after day after day. And you can plot these results, these two numbers—the angle from north, the angle up from the horizon. You can plot them day-by-day on a piece of graph paper, and you'll see the repeating pattern of the Moon. Note it's phase also, because the Moon will change from new to half, to full, to half, and so forth; day-by-day, you'll see those changes. This simple exercise repeats what has been done for thousands of years by ancient astronomers all over the world. It's a very simple observation that, in spite of our modern technological society, it's still worth doing today.

Some ancient societies displayed their belief in the predictability of the heavens by constructing large monuments in which the position of massive stones were aligned with key events in the sky. One of the earliest examples of human faith in this regularity lies at New Grange, Ireland, about 40 miles north of Dublin. There you can find the remarkable 5,000-year-old burial tomb of an ancient Irish leader. The tomb is in the form of a burial mound. It is constructed from a huge pile of heavy stones, and the burial chamber itself is 69 feet deep inside this mound. At dawn, on the winter solstice, that's the shortest day of the year when the Sun finally begins to track back after the days have gotten shorter and shorter and shorter, you see a long narrow channel and that shortest day the Sun's rays enter. Right at sunrise the Sun's rays enter that long channel, and they shine down 69 feet onto a rock with rock carvings right at the very center of the burial chamber. To build that chamber required tremendous effort and ingenuity, and it required faith that year after year after year the Sun would rise at a certain point on a certain day.

Stonehenge is another example. It's a great monument on England's Salisbury Plain in southern England. It served as an elaborate and sophisticated calendar. The agricultural community that labored to erect this great mass of stone—the great trilithons with two verticals and a horizontal lintel that weighed many tons—that great effort served to sight certain key times of the year: sunrise, sunset, moonrise, and moonset on certain key days. And you could, by sighting through the arches at more distant stones, at wooden pillars that were put around it at various points at that monument, you could basically keep time in the year. It was an astronomical calendar. There is a wonderful book about this subject, about Stonehenge; it is Richard [sic Gerald] Hawkings's *Stonehenge Decoded*. And in this book, Hawkings presents the theory that they could actually produce lunar and solar eclipses using Stonehenge. It's a fascinating idea, and certainly it's well worth the read if you're interested in those additional aspects of Stonehenge. But in any case, Stonehenge shows that ancient people recognized regularities and predictabilities in nature.

Science depends on maintaining meticulous records of observations, and so we're going to fast forward now to the time of Gaius Plinus Secundus, who lived from 23 to 79 A.D. He is best known as Pliny the Elder, and that's to distinguish him from Pliny the Younger, who was his nephew. Pliny the Elder led a rich and varied life. He was a cavalry officer in Rome's colonies in Germany; he served as a naval commander in Naples. For a time he was a lawyer. He also was an administrator to Rome's Spanish colonies, so he really got around ancient Rome. He's best remembered for his encyclopedic *Natural History*—a compendium that is divided into 37 books, each of which tabulates the facts that were known to him directly, or through respected authorities of the time. Topics ranged from a systematic catalogue of all the known plants and animals and minerals to various recipes for medicines and for manufacturing processes. This *Natural History* is quite unusual and it is the first book that actually cites references. It has over 100 references to 100 noted authorities.

This work provides an unparalleled view of knowledge at the time of the early Christian era, but the result is often an uncritical mix of facts and myths—superstitions. Pliny's translator, Professor Rackham of Christ College Cambridge, describes the work as follows: He says, "Pliny claims in the preface that the work deals with 20,000 matters of importance drawn from 100 selected authors. In selecting from these, he has shown scanty judgment and discrimination, including the false with the true at random. His selection is colored by his love of the marvelous, by his low esteem of human ability, and his consciousness of human wickedness. In his section on gem stones, for example, Pliny claims the quartz crystals form from intense freezing of moisture from the sky. Amber, he says, is a prophylactic against tonsillitis and other afflictions. And diamonds can be fractured by soaking them in goat's blood." There are also many facts that we would now counter as true, so there's a mixture in there. Pliny died, by the way, while observing an eruption of Mount Vesuvius in A.D. 79. This is the same eruption that destroyed Pompeii and Herculaneum. He first approached the volcanic eruption out of curiosity; he wanted to see what was going on. But when he got near, he got trapped because the seas became violent, there were earthquakes, and he died of asphyxiation while trying to comfort other victims of this disaster. Pliny's work, his *Natural History*, represents an essential step, and essential aspect of science—collecting data. It really wasn't the scientific method, so I want to look now at one of the first clear examples of the scientific method in action.

To many ancient cultures, knowing the positions of the Moon and the planets was of great importance, both for astrological predictions and for navigation—those two reasons why you'd want to study the skies. The thousands of visible stars formed this constantly rotating backdrop of constellations, and against those constellations you'd see night by night, the movement of planets, the wanderers. One of the most complicated things about the movement of planets is that they have retrograde motion. If you plot the position of the planets against the stars night after night, you'll see the planets moving forward, then they sort of slow down and for several weeks or months they move backward against the drop, then they start moving forward again, but then they slow down and start moving backwards. This retrograde motion had to be explained by any model of the universe.

One of the first descriptions of the universe that led to testable predictions was the extremely successful model of Ptolemy of Alexandria, who lived from 100 to 170 A.D. He was an Egyptian-born Greek. He collected many of his own observations about the heavens as well as compiling those of earlier Babylonian and Greek astronomers. He compiled all his information in a 13-volume work called the almagest. It's a lengthy treatise and it's best remembered for the Ptolemaic model of the solar system. Ptolemy argued that the Earth had to be at the center of the solar system, because otherwise if the Earth moved around some other object, the atmosphere would be stripped away. The Ptolemaic model postulated perfect circular orbits, but inside those orbits were smaller circles called epicycles, which explained the retrograde motion. This use of circles was absolutely dictated by the idea that the heavens were perfect, and the circle was the only perfect form.

Milton, the great poet, captured the complexity of this system in famous lines from his *Paradise Lost*. Let me read you this quote, "How they will weild the mightie frame, how build, unbuild, contrive to save appeerances, how gird the Sphear with Centric and Eccentric scribl'd o're, Cycle and Epicycle, Orb in Orb." This gives you a sense of the artificial complexity of this model. Yet in spite of its complexity, the Ptolemaic system was reasonably successful in predicting planetary positions, and it survived for 1,400 years; that's probably the longest running scientific theory in history. You could, using this Ptolemaic method, and extremely laborious mathematical calculations, predict the position of a planet at some future time and place. It was not until the 1500s that a new, more precise set of observations in the application of the scientific method led to a general acceptance of a competing theory.

In summary, this lecture we've seen that the scientific method is a richly varied human construct. It seldom operates as a simple cycle. Human intuition and imagination, creativity all play central and vital roles in the scientific method. The idea of a regular and predicable universe with overarching laws to be discovered by observation, and by experiment was not at all obvious to ancient societies. Aristotelian philosophers favored pure logic over experiments and understanding the world. Nevertheless, mathematical marvels of the solar system notably the Ptolemaic Earth-centered model displayed a faith in the regularity of nature. In the next lecture, we're going to see how the Ptolemaic view was eventually replaced by the Copernican model.

Lecture Four
Celestial and Terrestrial Mechanics

Principle: *"Empirical mathematical laws describe motions in the heavens and on earth."*

Scope:

This lecture introduces individuals who played pivotal roles in the history of 15th- and 16th-century science. Polish astronomer Nicolas Copernicus (1473–1543) devoted much of his life to developing a mathematical model of the solar system in which the Earth and other planets orbit the Sun—an alternative hypothesis to Ptolemy's Earth-centered system. Danish astronomer Tycho Brahe (1546–1601) advanced the field principally by designing and constructing greatly improved instruments that increased the precision and accuracy of astronomical observations. Tycho's meticulous observations revealed errors in the predictions of both the Ptolemaic and Copernican systems.

Upon his death, Tycho's wealth of data came into the hands of his mathematically gifted assistant, Johannes Kepler (1571–1630). Analyzing the data for Mars, Kepler derived three laws of planetary motion, including the idea that planetary orbits are elliptical with the Sun at one focus.

Galileo Galilei (1564–1642) transformed both the content and the methodology of science. He pioneered the use of the telescope in astronomy, and his published descriptions of "imperfections" in the heavens ultimately led to his heresy trial in 1633. But to many scientists, Galileo is honored first and foremost as the founder of experimental science.

Outline

I. Science is a collective human endeavor, advanced by thousands of little-known researchers who, step-by-step, add to the store of human knowledge. Yet the history of science often must focus on the seminal contributions of a few extraordinary individuals who played pivotal roles in synthesizing previous ideas. This lecture introduces four key scientific figures of 15th- and 16th-

century Europe: Nicolas Copernicus, Tycho Brahe, Johannes Kepler, and Galileo Galilei.

II. Polish astronomer Nicolas Copernicus (1473–1543) was trained in theology and spent nearly half a century working for the Catholic church. Yet, for reasons that remain uncertain, he devoted much of his life to constructing a mathematical model of the solar system in which the Earth and other planets orbit the Sun—a rival hypothesis to the prevailing Ptolemaic Earth-centered model.

A. Copernicus never sought personal recognition for this theoretical scholarly effort, and the model remained unpublished until 1543, the year of his death.

B. In his great work, *On the Revolutions of the Celestial Spheres*, Copernicus proposed the modern model, with the Earth and other planets orbiting the Sun.

1. Copernicus noted that for the stars to orbit the Earth, they would have to travel at enormous speeds.
2. The model explained the retrograde motion of Mars and other planets, which was a consequence of the Earth swinging from one side of the Sun to the other during its orbit around the Sun.
3. The model still relied on perfect circular orbits and small epicycles.
4. The Copernican model led to greatly improved predictions of planetary positions, which led to its acceptance.
5. The year 1543 also saw the publication of the seminal work in human anatomy, *On the Fabric of the Human Body*, by the great Flemish physician Andreas Vesalius (1514–1564), with magnificent illustrations crafted in Titian's studio.

III. The Copernican model of the solar system had to be tested, and that task fell principally to Danish astronomer Tycho Brahe (1546–1601). Tycho, as he is known, advanced the field by designing and constructing greatly improved astronomical instruments.

A. Tycho led a remarkable life. Abducted in childhood by his childless uncle, he received the best possible education and rose to great fame.

1. Tycho first came into prominence in 1572 at the age of 27, when he discovered a new bright star—a supernova in the constellation Cassiopoea.
2. King Frederick II of Denmark and Norway rewarded him by giving him title to an island between Denmark and Sweden, and building him a castle and observatory.

B. Tycho Brahe's improved instruments allowed him to test the predictions of both the Ptolemaic and Copernican systems.
1. Tycho's newly designed instruments reduced observational errors by a factor of 20.
2. Tycho revealed discrepancies in the predictions of both the Ptolemaic and Copernican systems.

IV. Upon his death, Tycho Brahe's data came into the hands of his mathematically gifted assistant, Johannes Kepler (1571–1630).

A. Kepler had become Tycho's assistant in 1600, and he carried on the work of observation and analyses that his mentor had begun.

B. Kepler derived three empirical laws of planetary motion that provided a mathematical description of solar system orbits:
1. Planetary orbits are elliptical with the Sun at one focus.
2. An imaginary line from a planet to the Sun sweeps out equal areas in equal times.
3. The square of a planet's orbital period (its "year") is proportional to the cube of its mean distance from the Sun.
4. These laws placed the Copernican model on a firm mathematical footing.

C. In introducing these ideas, Kepler expressed his deep conviction that the universe holds a deep order. Yet, in spite of the mathematical logic and order that Kepler brought to the heavens, he still had to contend with a superstitious world. In 1615 his elderly mother was jailed and brought to trial for witchcraft after being accused by a vindictive neighbor.

V. Galileo Galilei (1564–1642) transformed both the content and the methodology of science. He made major contributions to the fields of astronomy and physics, devised several ingenious practical inventions, and was a founder of experimental science.

A. Galileo, son of a Florentine musician, began his schooling in medicine, but soon was drawn to mathematics and natural philosophy. He was a brilliant thinker, but could be tactless and arrogant, and thus made many enemies, both in academic circles and among the leading politicians of the day.

B. Galileo pioneered use of the telescope, which he used to reveal unanticipated "imperfections" in the heavens. His first observations were published in *The Starry Messenger* in 1610.
1. Throughout the book, Galileo emphasizes the importance of modern observations over ancient authority.
2. Galileo observed craters and mountains on the Moon, Saturn's rings, and sunspots, all of which challenged the prevailing view that celestial objects are perfect spheres.
3. He documented numerous stars not visible to the unaided eye, thus undermining the authority of Aristotle and other ancient scholars.
4. Galileo observed four moons of Jupiter, which demonstrated that not all celestial objects orbit the Earth.
5. Galileo's publication of these discoveries, and his bold support of the Copernican system in his *Dialogue Concerning Two World Systems* (1632), ultimately led to his famous and frequently oversimplified heresy trial in 1633.

C. To many scientists, Galileo is honored first and foremost as a founder of experimental science.
1. He sought to understand the mathematical laws that describe falling objects—an important aspect of terrestrial mechanics.
2. Aristotle said that heavier objects fall faster than lighter objects, a claim that Galileo demonstrated was false.
3. Realizing that free-falling objects move too fast to measure with the observation techniques of his day, Galileo devised an ingenious adjustable ramp to "dilute" the effects of gravity.
4. This is known as the "rolling ball" experiment, which demonstrates that the distance traveled by a falling body is proportional to the square of the time of the fall.

D. Galileo discovered that the horizontal motion of a falling object is independent of its vertical motion. He tested these ideas experimentally, by firing cannonballs off of a cliff and observing the curving path of the fall.
   1. Galileo's empirical laws of terrestrial motions, are not unlike those of Kepler for planetary motions.
   2. It remained for Isaac Newton to merge the empirical laws of Kepler and Galileo into one set of universal laws of forces and motions.

Essential Reading:
Trefil and Hazen, *The Sciences: An Integrated Approach*, Chapter 2.

Supplemental Reading:
Andrade, *Sir Isaac Newton*, Chapters 1–2.
Caspar, *Kepler 1571–1630*.
Cohen, *Birth of a New Physics*, Chapters 4–6.
Drake, *Galileo: Pioneer Scientist*.
Harre, *Great Scientific Experiments*, Chapter 6.
Kuhn, *The Copernican Revolution*.
Zeilik, *Astronomy,* Chapters 2–4.

Questions to Consider:
1. Based on the discoveries of Kepler and Galileo, how would you characterize the role that mathematics plays in science? Is the correspondence between the geometrical abstraction of quadratic equations (including ellipsoids and parabolas) and motions in the natural world purely coincidental?
2. If you were to duplicate Galileo's rolling ball experiment today, how might you improve on his measurements of distance and time? (Assume an unlimited research budget.)

Lecture Four—Transcript
Celestial and Terrestrial Mechanics

The intense effort to observe, to describe, to understand what's going on in the heavens was largely an empirical endeavor. Ancient astronomers looked into the heavens and saw what was there, but these early efforts went beyond a simple description. Indeed, these pioneer astronomers were amongst the very first people to recognize the power of mathematics in describing nature. It's a great principle of science that empirical mathematical laws describe motions in the heavens and on Earth.

Science is a collective human endeavor advanced by thousands of little-known researchers who collect data step-by-step, who record them, and who publish these data. You'll never hear about these forgotten observers. Relatively few individuals make the great intuitive leaps, and those are the individuals who we justly remember. Consequently, the history of science often must focus on the seminal contributions of a few key individuals, and ignore the roles of all of those other thousands of individuals who collected data. This is a fact of life in the history of science. We don't know the identities of the ancient astronomers who built Stonehenge or other monuments, but Greek and Roman scholars, such as Aristotle and Pliny, of Ptolemy—these are people who were celebrated in their own day, and they are remembered today as well.

My objective in this lecture is to introduce four key individuals who transformed our understanding of the heavens. These are individuals, figures of the 15th and 16th-century Europe. We're going to meet the Polish astronomer, Nicolas Copernicus, who presented a sun-centered model of the solar system. We're going to meet the Danish astronomer, Tycho Brahe, who dramatically improved observational instruments in astronomy; then the German mathematician, Johannes Kepler, who analyzed the geometrical characteristics of motions of planets in the heavens; and finally, the Italian physicist, Galileo Galilei, who was a pioneer in the development of the experimental method in science.

These scientists lived at a time when conventional scholarly wisdom valued ancient authority over experiment and observation. In this context, the Earth was seen as a place of decay, of change. It was placed at the fixed center of the universe where the immutable heavens surrounded it. The heavens were the place of perfection, circular orbits, fixed positions. All the objects in the heavens were assumed to be perfect spheres. As we shall see, it's the work of Copernicus and Tycho, of Kepler and Galileo, that changed the human perspective of our place in this universe.

Our first character is the Polish astronomer, Nicolas Copernicus, who lived from 1473 to 1543. He was trained in theology and spent nearly half a century working for the Catholic Church. Yet, for reasons that remain uncertain, he devoted much of his life to try to construct a mathematical model of the solar system, a model in which the Sun, not the Earth, was at the center. This was a rival hypothesis to the prevailing Ptolemaic Earth-centered model. It's hard to understand why Copernicus did this work. Why would someone who devoted his life to the church spend so much time, his spare time, developing a model that was against church doctrine—putting the Sun at the center rather than the Earth.

Copernicus seems to have been a very modest man. He studied at the University of Krakow in Poland, and he credited to the university throughout his writings for all of his intellectual accomplishments. He said, "It wasn't me, it was just what I learned at the university." He went to Italy to obtain training in medicine and in clerical law, simply so that he could help out his own clerical order, so he could serve the community better. He built his own astronomical observatory in his spare time. He made his own records of the positions of stars and the Moon. Although he's not best remembered as an observational astronomer, rather, as a theoretician who developed this new model.

A preliminary manuscript version of his heliocentric theory, the sun-centered theory, was circulated as early as 1514, and a few copies of that manuscript survive. Copernicus, however, did not sign his name to the theory, to this essay that he distributed to some friends. Indeed, he never sought personal recognition for this theory and the theory was not published until right around the time of his death in 1543. Now think about it, this is a really strange kind of hobby, this mathematical tinkering with the cosmos, seeing if you could fit the orbits of planets into this alternate view. His great work is called *On the Revolutions of the Celestial Spheres*. And in it, Copernicus proposed essentially the modern model of the solar system where the Earth and the other planets all circle the Sun, while the Moon circles around the Earth. This theory, of course, is in sharp contrast to the Ptolemaic view in which you had the Earth at the center, and the planets circling the Earth, with their circular orbits, but also the epicycles to explain retrograde motion. Copernicus noted that for the stars to orbit the Earth, they'd have to be traveling at an enormous speed. Far simpler, he thought, for the Earth to rotate on its axis once every day and that each year represents one orbit of the Earth around the Sun, and that's the modern view.

The Copernican model explained most of the retrograde motion of planets like Mars and so forth. And he said that was merely a consequence of the fact that the Earth swings in its orbit from side to side around the Sun, and as it swings back and forth, you see Mars against the backdrop of the stars also moving back, and that explains the retrograde motion of Mars. But the model still relied on perfect circular orbits, because Copernicus was clearly in that tradition, and so he still had to use epicycles, although much smaller than the epicycles of Ptolemy. The Copernican model led to a greatly simplified and more accurate prediction of planetary positions, and that led to its acceptance by many scholars. That, after all, is the true measure of a theory, whether it works, whether it leads to useful predictions. That year, 1543, that's a very pivotal year in science. It also saw the publication of a seminal work on human anatomy, that is, *On the Fabric of the Human Body* by the great Flemish physician Andreas Vesalius, who lived from 1514 to 1564. That magnificent volume contains illustrations that were crafted in Titian's studio. It's a wonderful work if you've ever had an opportunity to see the illustrations from *On the Fabric of the Human Body*. The year 1543, consequently, is often cited as the dawn of the modern age of science.

The Copernican model of the solar system had to be tested, and that task fell principally to the Danish astronomer Tycho Brahe, who lived from 1546 to 1601. Tycho, as he is usually called, advanced the field principally by designing and constructing greatly improved astronomical instruments. This tremendously increased the precision and the accuracy of measurements made on objects in the heavens. Let me tell you about Tycho's life because it was quite remarkable. He was one of 11 children. He was abducted in childhood by his childless uncle, who was a prominent naval officer named Jorgen Brahe. Jorgen Brahe provided the eager young student with the best possible education, and I suppose, given the fact that his parents had

11 children, they probably really didn't notice that he was gone all that much. I can imagine the Brahe household must have been quite a site.

In any case, Tycho rose to great fame at an early age, and he enjoyed the favors of the Danish king, King Frederick II, who lavishly supported his astronomical research. That royal support, however, did not last throughout Tycho's lifetime. Frederick II died in 1588. His successor, King Christian IV, did not support astronomical work, and it's partly because Tycho was not an easy person to get along with. It is said that he was haughty and arrogant with members of the royal family who tried to tour his observatory. So, Tycho's own arrogance may have contributed to his downfall. Ultimately, he was exiled to Prague, and that's where he died in 1601.

Well, the young Tycho first came into prominence in 1572 at the age of 27 when he discovered a new bright star, a super nova, which is an exploding star. He discovered this in the constellation Cassiopeia, and for 18 months he detailed the gradual dimming of that star. These observations demonstrated the then astonishing idea that Aristotle didn't know everything there was to know about the heavens, and that the heavens themselves are not immutable, that they can change from year to year, and this added new impetus to the idea you really should study the heavens. This should be an ongoing process as opposed to just taking everything that Aristotle had said on faith. So King Frederick II of Denmark and Norway rewarded him by giving him title to an island, the Island of Hven, which is located between Denmark and Sweden. On that island the king had built a great castle and observatory, as well as workshops for the construction of the new equipment that Tycho designed. He spent more than 20 years on the island constructing these astronomical instruments and making various astronomical observations.

Tycho Brahe's principal task was in improving astronomical observations with new instruments and then using those instruments to test the Ptolemaic and the Copernican system. He designed devices called quadrants and sextants. These were devices that would locate the position of stars and planets in the heavens, and they were quite different from modern telescopes. They didn't employ lenses. They were just basically glorified gun sights, but very beautifully crafted gun sights. So you'd measure an angle around from the horizon and an angle up and sight a particular object at a particular time and place. And this is just like the exercise I described in the last lecture where you use a compass needle to get a compass bearing along the horizon. So you use your compass to get one angle, and then you use a protractor as a sighting device to get a second angle. What Tycho did was the exact same thing, but with very elaborate equipment, which were adjusted for thermal contraction of the cold night air, for example. He had very accurate timepieces to make sure he knew exactly when he was making his measurements. This is what he did: He improved the measurements so that before Tycho you could measure an angle of arc about 10 minutes, which is like having a dime at perhaps 20 feet away from you, and that's the plus or minus, that's the error of measurement. By the time that Tycho Brahe was finished with his new instruments, you could measure the position of a dime at more than 100 yards away. So imagine this dime 100 yards away, and still being able to measure its position exactly. That's the kind of increase that Tycho sought.

And so what he did is he'd look first at the Copernican predictions and then at the Ptolemaic predictions, for example, for the position of Mars at a specific time on a specific date. You find that position, and there's where Copernicus said Mars should be, and Ptolemy said, "No it should be over here." Tycho looks, and it's neither. The position of Mars is someplace nearby, but not exactly where either of the previous models said it should be. So what does Tycho do? Just what any good astronomer would do, he came up with his own model, which is a rather complex combination of Copernican and Ptolemaic ideas. Tycho's story reveals why improved instruments are so important in science. You can come up with a theory, but you have to make predictions, then you have to test those theories. Often, it's an improved instrument with greatly reduced errors that provides new insight.

And now we come to the third character in our cast. Upon his death in 1601, Tycho Brahe's data fell into the hands of his mathematically gifted assistant, the German mathematician Johannes Kepler, who lived from 1571 to 1630. Kepler cannot have had an easy childhood. His father was described as "criminally inclined, quarrelsome, and liable to a bad end." His mother was called "garrulous and bad tempered." Nevertheless, he was an excellent student. He excelled in math and in astronomy. Perhaps in the abstractness of these studies he escaped the turmoil of his family life.

Kepler had become Tycho Brahe's assistant in about 1600, and he carried on that work of observation at the observatories. He kept making measurements of Mar and other planets, for example. But he had an advantage. He was a mathematician, and he was able to take those numbers and fit the path of the planets to a mathematical equation, and this is what we'll see as Kepler's great advantage. He saw a great harmony, a great order to the natural world, and he derived three great laws of planetary motion. The first of these laws describes the shape of planetary orbits. Kepler says the planetary orbits are elliptical, with the Sun at one focus. Though I want to remind you of the shape in ellipse and how it's drawn with two foci and a curved path. And so I'm going to do a demonstration now to give you an example of how you can draw an ellipse at home and see how it differs from a circle. You can draw an ellipse at home by taking a board, hammering two nails through it, and then taking a loop of string. All you have to do is just keep that loop taut, as you go around the nails, you form that curving path, sort of like an elongated circle, and that's the ellipse. What Kepler said was that every planetary orbit is an ellipse, with the Sun at one focus. That's his first law.

His second law is that the speed that the planet travels as it's going around the Sun isn't constant. When you're close to the Sun, you travel very fast. When you're far away, you travel more slowly. So the equal areas are swept out in equal times, and you've probably seen this idea of the slingshot effect. A spaceship is coming closer and closer to the Sun. It speeds up and goes faster and faster, and it whips around and then comes back. That's the slingshot effect, and it has to do with Kepler's second law.

The third law is a mathematical relationship of exactly how fast a planet travels depending on its distance. Kepler's third law is as follows: He said, the square of a planet's orbital period—that is, the planet's year squared—is proportional to the cube of the average distance it takes from the Sun. So you can see how these laws have placed the Copernican model on a firm mathematical footing.

Let me give you an example about how you can use this law mathematically. Let's calculate the distance to Jupiter just by knowing its orbital period, which is 12 years, because you can make the observation that it takes Jupiter 12 years to travel from one part of the zodiac, all the way around back to the same part of the zodiac.

And here's how the calculation is done. According to Kepler's third law, the period squared, P^2, over the average distance cubed, or D^3, for Earth is equal to P^2 over D^3 for Jupiter. For the Earth it's easy. The period is one year. So, 1^2 is 1. The distance is one astronomical unit, so 1^3 is also 1. For the Earth the number is just 1. And that equals, 12^2—the orbital period for Jupiter squared, 12^2—or 144 divided by the distance cubed, and we have to solve for D here. So D cubed is equal to 144. That means the distance to Jupiter is the cubed root of 144, or 5.2 astronomical units, 5.2 times the distance from the Sun to the Earth. That was a pretty useful thing to know, and it fell right out of Kepler's laws.

In introducing these ideas, Kepler expressed his firm belief that the universe held a great deal of order, indeed, mathematical harmonies. In the introduction to his 1619 work called *Harmony of the World*, which introduced the third law, he says, "At last I have found it, and my hopes and expectations are proven to be true that natural harmonies are present in the heavenly movements—both in their totality and in detail—though not in a matter which I had previously imagined, but in another, more perfect manner."

In spite of this mathematical logic—an order that Kepler brought to the universe—he still had to contend with a superstitious world. In 1615, his elderly mother was jailed. And even as he was writing *Harmony of the World*, he had to go to his mother's hometown and testify and defend her at her trial for witchcraft. At that time in Germany, dozens of people had been tortured and executed for the offense of witchcraft, and so for six years Kepler was forced repeatedly to interceded on her behalf, even at great risk to his own well-being.

We now come to the fourth of these famous dead white males. Galileo Galilei lived from 1564 to 1642. He's one of a handful of people who transformed both the content of science and the methodology of science. He made major contributions to the fields of astronomy and physics. He devised several ingenious practical inventions, and was the founder of experimental science. He was the son of a Florentine musician. He was born in 1564, the same year as Michelangelo's death and Shakespeare's birth. He began his schooling in medicine, but soon was drawn to the beauty of mathematics and natural philosophy. Natural philosophy was the word for science back in those days. He was a brilliant thinker, but he could be tactless and arrogant. He thus made many enemies, both among Italian academics and also political and religious leaders at the time, not a good thing to do in Italy at that time.

Many people remember Galileo for his pioneering use of the telescope, which he used to reveal unanticipated imperfections, as they were called, in the heavens. He built his first telescope in 1609, first a nine-power instrument, then later a 30-power instrument, after he heard about the invention in Holland. In Holland it was just used as a curio, but Galileo realized this could be an important scientific instrument. His first observations were published in a book called *The Starry Messenger* in 1610. Throughout the book Galileo emphasizes the importance of modern observations over ancient authority. He promises to reveal "great, unusual and remarkable spectacles opening these to the consideration of every man, and especially philosophers and astronomers." This rhetorical challenge looked at the prevailing Aristotle view and said, "Aristotle might be wrong. He might not know everything there is to know about the heavens. We have to look for ourselves."

Galileo was the first to observe the craters in the mountains of the Moon—imperfections. He saw the compound nature of Saturn, what we now realize are rings. He saw the moons of Jupiter. He saw the phases of Venus. He documented thousands of stars that weren't visible to the unaided eye. He saw 80 new stars in Orion's belt alone. He realized that the Milky Way is a band of countless thousands of stars that you can't see with your unaided eye. Again, these findings undermined the authority of Aristotle, because it showed that Aristotle didn't know everything there was to know. You can make all of these observations yourself with a good pair of binoculars, a 10-power pair of binoculars you can see the moons of Jupiter, you can see the rings of Saturn. It's really worth doing sometime if you've never done that because they are beautiful sights. You can make the same observations that Galileo made.

Now, he was not a mere describer of the astronomical features. Historian of science, I. Bernard Cohen describes the power of Galileo's work as follows: "Not only did Galileo describe the appearance of mountains on the Moon, but he also measured them. It is characteristic of Galileo as a scientist of the modern school that as soon as he found any kind of phenomenon, he wanted to measure it." It is all very well to be told that the telescope discloses that there are mountains on the Moon, just as there are mountains on Earth. But how much more extraordinary it is, and how much more convincing to be told that there are mountains on the Moon and that they are exactly four miles high. Galileo's publication of these discoveries and his bold support of the controversial Copernican system and his dialogue concerning two worlds that was published in 1632, ultimately led to his famous and frequently oversimplified heresy trial, which took place in 1633.

Galileo had earlier promised church officials that he would not advocate the Copernican system, at least not in public discourse. But in that 1632 book, which was published in Italian, not in Latin, which is the scholarly language, he supposedly presents an even-handed account of both this Earth- and the sun-centered view. But it really wasn't an even-handed account. The book adopts the literary form of a debate, a conversation among observers of Copernican and Ptolemaic leanings, with an educated layperson sort of sitting around asking questions of these people. Galileo remains, at least in a literal sense, uncommitted, but he puts the arguments of the Ptolemaic viewpoint in the mouth of a narrow-minded Aristotelian and someone who bears resemblance, both physically and in terms of rhetorical approach, to Pope Urban VIII, who was the Pope at the time. Galileo's readers knew exactly what his point of view was, and the Pope took great offense, as is not a big surprise. So Galileo was convicted of heresy. He was forced to recant the Copernican view, and he was placed under house arrest for the rest of his life, the final decade of his life he remained at home until his death in 1642. By the way, the Catholic Church officially reopened that case in 1992, and exonerated Galileo 350 years after his death.

To many scientists, Galileo was honored first and foremost as the founder of experimental science. Along with many of his predecessors and contemporaries, he sought to understand the mathematical forms, the laws that described falling objects. This is extremely important in an age when cannons had just been developed (and gunpowder and explosives), and you need to be able to fire objects accurately from one place to another. You needed to know what sort of curving paths objects adopted when they were fired in the Earth's gravity. Aristotle had said that heavier objects fall faster than light objects, and this is a claim that Galileo demonstrated that is quite false. The story that Galileo dropped two balls from the Leaning Tower of Pisa, by the way, is probably apocryphal, but nevertheless he did do that kind of experiment where he'd take two objects of different masses and different sizes and drop them from a high place, and find that they landed at exactly the same time. So, through an empirical experimental approach, he showed that the reasoning, the rationale of Aristotle was wrong. Realizing that free falling objects move too fast to measure with any sort of conventional techniques of the day—the watches and clocks that they had available at that point—Galileo devised an ingenious adjustable ramp to dilute the effects of gravity. This is called the rolling ball experiment, and I'd like to share that with you in just a moment.

The main problem with the "rolling ball" experiment is that you need accurate time measurements. In Galileo's day, there weren't really any accurate timepieces. You didn't have watches. You didn't have stopwatches, or clocks of that sort. At first, Galileo used his pulse, but that wasn't very accurate. Then he invented an ingenious way to measure time. He said, "We employed a large vessel of water and placed it in an elevated position. To the bottom of this vessel was soldered a pipe of small diameter, giving a thin jet of water. We collected this water in a small glass during the time of each descent. The water thus collected was weighed after each descent on a very accurate balance. The differences and ratios of these weights gave us the differences and ratios of the times." Galileo and his assistants conducted numerous repetitions—another aspect of experimental science. Let me read you another quote. "In such experiments repeated a full 100 times we always found that the spaces traversed were to each other as the squares of the times, and this was true for all inclinations of the channel along which we rolled the ball."

Let me do just a couple of examples of this rolling ball experiment to show you what Galileo did several hundred years ago. Galileo realized that objects fall straight down much too fast to record how they fall, so he invented an experiment—an inclined plane—which diluted the force of gravity. What he'd do is measure a distance along the inclined plane, and then time the fall. Zero, one, two, three, four, five, six. It takes six units of time to go the entire length, and you might guess, well then it would only take three units to go half as far marked here. Let's try that. Zero, one, two, three, four. But it takes four units, not three. You have to go all the way down here, to only a quarter of the original length. And here—zero, one, two, three. Three units to travel one quarter of the distance, and six units to travel the entire distance. So distance is proportional to the square of time. That was Galileo's great discovery with the rolling ball experiment.

A fascinating aspect of this experiment is that Galileo did not conduct the rolling ball experiment to discover a mathematical relationship between time and distance. Rather, he used the apparatus to confirm his conviction that velocity and time bear the simplest kind of relationship to each other. That is, the velocity of a falling object is proportional to the time of its fall. He called this steadily increasing velocity, uniform acceleration. Galileo also demonstrated mathematically that this result was equivalent to saying that the distance traveled by a falling object is equal to the square of the time of its fall.

Galileo devised lots of other experiments in his study of terrestrial mechanics, and you can repeat some of these yourself. One of his most famous is his study of pendulums. I'm not going to go through all of his apparatus, but you can do this at home. You can take a stopwatch. You can have shorter pendulums. You can have longer ones. You can put different weight and masses on them and see what happens. What you'll find is that longer pendulums swing more slowly than short ones, but that the rate of speed is independent of the mass of the pendulum. That's the sort of thing that's fun to do at home, and try various objects with your kids if you can. It's a great thing to do at home.

Galileo also discovered a key principle regarding ballistics, that is, the way objects fly through the air. He found that horizontal motion of a falling object is completely independent of the vertical fall. For example, he cited the example of a heavy object dropped from the mast of a moving ship. Aristotelian philosophers held that an object would land some distance behind the mast of the moving ship if you dropped it high up the object would fall backwards, if you will, and the ship would move out from under it. But Galileo said no, the objects moving along with the ship is going to fall right at the base of the mast. He tested these ideas experimentally, for example, by firing cannonballs horizontally off a cliff, and observing the curving path of the fall. And what he found that is when you do that, you come up with a curved path, called a parabola, and that curving path you see over and over again, that's the curve of falling objects.

To summarize this lecture, we've met four key scientific figures of the 15th and 16th centuries. We've seen Polish astronomer Nicholas Copernicus, who developed a model of the solar system in which the Earth orbits the Sun; Danish astronomer Tycho Brahe, who tested the Copernican model with dramatically improved astronomical instruments; the German mathematician Johannes Kepler, who derived new mathematical laws for the movement of planets in their orbits; and finally, the Italian physicist Galileo Galilei, who developed a mathematical description of motions of objects at the Earth's surface. Galileo's empirical laws of terrestrial motions, which described curved paths of moving objects near the Earth and the mathematical relationships of distances and velocity in time, are not unlike those of Kepler in his descriptions of the orbits of planets. But it remained for Isaac Newton to merge the empirical laws of Kepler and Galileo into one set of universal laws of motion and that's the subject of the next lecture.

Lecture Five
Newton's Laws of Motion

Principle: *"One set of mathematical laws, formulated by Isaac Newton, can be used to predict the motion of objects anywhere in the universe."*

Scope:

By 1640, Kepler had derived his laws of planetary orbits and Galileo Galilei had presented equations that predicted the behavior of falling objects. Yet, at that time, the studies of terrestrial and celestial mechanics remained separate domains. Isaac Newton (1642–1726) synthesized empirical descriptions of celestial and terrestrial motions into one set of laws that applies to motions anywhere in the universe. During a remarkable period in 1665–1666, he formulated many of his major contributions to science, including the branch of mathematics called calculus, many of the basic laws of optics, the laws of motion, and the law of gravity.

Newton divided all physical movement into two categories. Uniform motion is the movement displayed by objects traveling in one direction at a constant speed, whereas acceleration is any motion involving a change in direction or speed. The first law of motion states that any object in uniform motion will remain in that motion unless acted upon by a force. The second law of motion defines an exact mathematical relationship: force equals mass times acceleration. The third law of motion is the familiar but subtle statement that for every action there is an equal and opposite reaction. Newton's three laws of motion together provide a complete framework for investigating all forces and motions that occur in our lives.

Outline

I. By 1640, Johannes Kepler had derived the mathematical foundation for describing planetary orbits around the Sun, and Galileo Galilei had presented equations that predicted the behavior of falling objects. Yet, at that time, the studies of terrestrial and celestial mechanics remained separate domains, presumably with quite different sets of natural laws. It remained for Isaac Newton (1642–1726) to synthesize the empirical descriptions of celestial and terrestrial motions into one set of laws that applies to motions everywhere in the universe.

- **A.** Newton was a student of mathematics and natural philosophy at Cambridge University, from which he graduated with a Bachelor's degree in 1665. In that same year, before he could return for further studies, the Great Plague struck England.
- **B.** With the University shut down, Newton spent 18 months at his family farm in intense personal study and thought. During that remarkable year and a half he formulated many of his major contributions to science, including the branches of mathematics now known as integral and differential calculus, many of the laws of optics, the universal laws of motion, and the law of gravitation!
- **C.** Newton returned to Cambridge University in 1667, and remained at Trinity College for the rest of his life.

II. In formulating his three laws of motion, Newton first divided all physical movement into one of two categories.
- **A.** Uniform motion is the movement displayed by objects traveling in one direction at a constant speed.
    1. Uniform motion includes the special case of an object at rest.
    2. Previous scholars had also considered the circular orbits of planets a kind of uniform motion.
- **B.** Newton classified orbital motions incorporating a change in direction, as well as any motion involving a change in speed, to fall in the second category of motion, known as acceleration.
    1. Acceleration includes both speeding up and slowing down.
    2. Acceleration also occurs any time direction is changed, whether or not the speed changes.

III. Newton's first law of motion states: "Every body continues in its state of rest, or of uniform motion in a [straight] line, unless it is compelled to change that state by forces impressed upon it."
- **A.** Objects can display three distinctive types of behavior.

1. An object can move at constant speed and direction, varying in neither speed nor direction.
2. An object can stand still.
3. An object can accelerate under the influence of a net force.

B. The first law of motion provides an operational definition of "force" as the phenomenon that causes an object to accelerate.

C. The idea that circular planetary orbits, as well as all other accelerations, are caused by an unbalanced force was radical in Newton's day, for it demanded that forces act over the huge distances of space.

IV. Newton's second law of motion defines the exact mathematical relationship among three measurable physical quantities: force, mass, and acceleration. "The acceleration produced on a body by a force is proportional to the magnitude of the force and inversely proportional to the mass of the object."

A. Because mass is measured in kilograms (kg) and acceleration is measured in meters per second per second (m/s2), the unit of force is the kg-m/s2, called the "Newton." Mass in this equation represents an object's tendency to resist acceleration—a property called its "inertia." The more massive an object, the greater the force necessary to impose a given acceleration, and thus the greater its inertia.

V. The third law of motion is the familiar but subtle statement: "For every action there is an equal and opposite reaction." Forces always act simultaneously in precisely balanced pairs.

A. If you apply a force in throwing a ball, the ball applies an equal and opposite force to your hand.

B. Newton's three laws of motion together provide a complete framework for investigating and understanding all of the forces and motions that occur in our lives. Newton, himself, applied this framework to describe the ever-present force of gravity.

Essential Reading:
Trefil and Hazen, *The Sciences: An Integrated Approach*, Chapter 2.

Supplemental Reading:

Andrade, *Sir Isaac Newton*, Chapter 3.

Cohen, *The Birth of a New Physics*, Chapter 7.

Questions to Consider:

1. Identify specific ways in which Newton's three laws of motion have come into play in your life during the past 5 minutes.
2. What forces besides gravity affect you in your daily life? In other words, what forces besides gravity cause objects to accelerate?

Lecture Five—Transcript
Newton's Laws of Motion

The power of science lies in its ability to make connections between seemingly unrelated phenomena. And what could seem more unrelated than the motions of objects at the surface of the Earth—the way things fall, the way things fly through the air—compared to the stately motions of planets and stars in the heavens. How astonishing then, that one set of mathematical laws formulated by Isaac Newton can be used to predict the motions of objects anywhere in the universe.

In this lecture, the fifth of our series, I want tell you about one of the giants in the history of science—Isaac Newton. It would be hard to overstate the influence of Newton on the progress of scientific understanding. He saw the cosmos in an entirely new way, and he changed the way science is done. My principal objective in this lecture is to introduce this remarkable figure and look in detail at his three universal laws of motion.

Before I introduce our lead actor, however, I want to briefly review the history of science leading up to Isaac Newton. Five thousand years ago there was no science, but ancient peoples recognized certain patterns in nature—certain predictable patterns in the position of the Sun and the Moon and the stars, and also in the way objects fall on the surface of the Earth. Carefully oriented monuments like Stonehenge are mute testimony to the skill and the beliefs of these ancient peoples. Subsequent scholars focused on the irregular wanderings of planets, and by the time of Ptolemy, in A.D. 100 or so, centuries of written records allowed him to propose a complex mathematical model of the solar system, one in which the Earth is at the center and the planets and the Sun orbit around the Earth. The Ptolemaic system was one of the longest lasting scientific models of all time.

The European Renaissance saw a tentative questioning of many of the previous ideas. Polish astronomer Nicolas Copernicus, for example, suggested an alternative to the Ptolemaic universe. His model of the solar system had the Sun at the center and the Earth orbiting around the Sun. What a surprise then, with these two conflicting models when Tycho Brahe tested both the Copernican and Ptolemaic models and found that neither one of them worked.

Both fell short, because neither made accurate predictions about the exact positions of planets. It fell to the mathematician Johannes Kepler then, to show that the previous assumptions of circular orbits were in error. Indeed, planets orbit around the Sun in elliptical orbits, orbits that are slightly elongated from a pure circle. And then we had Italian physicist Galileo Galilei. He used mathematical analysis in his study of terrestrial mechanics, of the falling of objects, of the rolling of objects down an inclined plane—a subject that was of great interest to military leaders at that time.

So, we now find ourselves in the 1640s. Johannes Kepler has derived the mathematical foundation for describing planetary motions, Galileo has presented empirical, mathematical relationships for how objects fall. And yet, at that time, these studies of terrestrial mechanics and celestial mechanics were quite separate domains because people did not see the motion of objects in the heavens as at all related to those on the surface of the mundane Earth. It remained for Isaac Newton to synthesize these empirical descriptions of celestial and terrestrial mechanics into one set of laws that applies to motion everywhere in the universe.

Let me tell you a little bit about the life of this remarkable man. Newton experienced quite a difficult childhood. He was born prematurely on Christmas Day in 1642. That was about three months after his father's death. He was so tiny at birth the doctors didn't think he was going to survive, but he eventually thrived as a young man. He grew up in a family home at Woolsthorpe, England, that's a small village in Lincolnshire in east central England. When young Isaac was only three years old, he was abandoned by his mother, who remarried a minister and moved away. Newton then lived with his grandmother until the age of 11, and that's when his stepfather died. His mother returned to Woolsthorpe, that was in 1653, with her three younger children by the second marriage. And can you imagine the resentment and the feelings of abandonment that Isaac Newton must have felt at that point as a twenty-two-year-old child, with his mother just returning for the first time?

As a child, it is said that Newton loved mechanical devices. He was said to have constructed his own clocks of ingenious design. He said that he made a working model of a mill and used a small mouse to power that mill. He also enjoyed drawing, and he attended schools continuously from the age of five. But upon her return to the family farm, his mother took Isaac out of the school that he loved and forced him as a teenager to tend to the Lincolnshire farm. This was much against his wishes, but by all accounts, Isaac Newton was a terrible farmer. He was a failure. He neglected his chores, and he was absentminded in the whole business of farming. As a result, family friends encouraged his mother to let Isaac return to school, and this is a great good fortune for all of us in science, because after additional studies, he was admitted to Trinity College in Cambridge (that was in 1661). He focused primarily on mathematics and in natural philosophy, and then he graduated with a bachelor's degree in 1665. It turns out that while he was in school, in addition to many, many classical studies by the classical scholars, Newton is known also to have studied the writings of Kepler and of Galileo during those critical years when he was an undergraduate.

And here we come to a critical, pivotal year, 1665. In that year, before he could return to Cambridge for additional studies, the Great Plague struck England. The effects of the bubonic plague in England were devastating, killing perhaps one out of every seven people living in London during those periods. Cambridge University simply closed down and sent its students home because population centers were most likely to be struck by this devastating disease. So with the university shut down, Newton spent 18 months at the family farm in intense personal study and thought. During that remarkable year and a half, he formulated four major contributions to science and mathematics. Any one of these would have assured a lasting place in the history of science. He did four of them in a year and a half. He developed the branch of mathematics now known as integral and differential calculus. He discovered many of the laws of optics, especially relating to the nature of white and colored light. He formulated his three universal laws of motion, which I'm going to describe in this lecture, and he derived the universal law of gravitation, and that's the subject of the next lecture. In Newton's own words, "All this was in the two plague years of 1665 and 1666, for in those days I was in my prime of age for invention, and minded mathematics and philosophy more than at any time since." Many of the ideas developed during those years weren't published until more than 20 years later, many of them in Newton's great work, the *Principia* of 1687. Well, Newton returned to Cambridge University in 1667, and he remained there at Trinity College as an increasingly revered scholar with higher and higher academic positions for the rest of his life.

Let me set the stage for Newton's laws of motion. Before trying to understand these laws of motion, it is important to categorize the different kinds of motion—that's the kind of pattern in nature. And here's the scientific method at work. You observe lots and lots of motion, and then you see patterns, and you come up with the hypothesis of the different kinds of motion. That's the first step. Previous scholars had treated planets and moons in circular orbits as being in uniform motion. That was distinct from different kinds of acceleration, such as uniform acceleration, which we saw in the rolling ball experiment. That's the acceleration of falling objects or the kinds of more rapid and varied acceleration you'd experience, for example, on a roller coaster ride.

In formulating his three laws of motion, Newton first divided all physical motion into two different categories of motion. But these differed from previous scholars. First of all he defined uniform motion as movement displayed by objects traveling in one direction at a constant speed, or the special case of an object at rest. So, if an object is just sitting there on a table, if it just sits there and stays still, it's in uniform motion; that's uniform motion. Also, uniform motion would involve taking an object, moving it in a straight line at constant speed, and as long as you move in a straight line at constant speed, you're in uniform motion. To Newton, every other kind of motion imaginable is called acceleration. Now this is in contrast to previous scholars who considered the circular orbit of planets as a kind of uniform motion. When an object moves like this in a perfect circle, at constant speed, previous scholars had said that's uniform motion, but not to Newton. He said orbital motion is a kind of acceleration because it incorporates a change in direction. Remember his definition: any change in direction, any change in speed, that's acceleration. Now remember, acceleration in Newton's definition includes both speeding up and also slowing down. So when you put the brakes on your car, according to Isaac Newton, you're actually accelerating. It's a negative acceleration, if you will. The easiest way to treat this concept mathematically is with something called a vector. In a vector in mathematics you define both a direction and a speed. This can be represented like a little arrow, where the arrow points in the direction, and the length of the arrow corresponds to the speed.

This distinction between uniform motion and acceleration was a critical advance of Isaac Newton; it was a great insight. It allowed him to formulate his three laws of motion, which prior to that time could not have been formulated. So let's go on now to Newton's first law of motion. It states, "Every body continues in its state of rest or of uniform motion in a [straight] line unless it is compelled to change that state by forces impressed upon it." In other words, you can observe three distinct types of behavior. An object can move at a constant speed, in a constant direction, varying in neither speed nor direction as it goes along. In real life we almost never see this kind of motion because there are forces all around us. You'd have to go to extremely deep space far away from any stars or planets to get a sense of this kind of uniform motion, but an object moving in deep space would approximate uniform motion of that sort.

You can also have an object standing still; that's a kind of uniform motion, and when an object stands still, it's not changing its velocity, it's not changing its direction, but it's the special case where that vector is just a point.

And then you can have an object that accelerates under the influence of a net force, and that's the key point in Newton's first law of motion—you can change the speed or you can change the direction, or you can change both under the influence of a force. And here is the breakthrough—Newton's first law defines force. It provides an operational definition of force. It's the phenomenon that causes an object to accelerate. Anytime an object accelerates, anytime anything interesting happens in the world around you, you might say, there has to be a force involved. This law, however, does not identify any specific force. In the coming lectures we're going to learn about several different forces—gravity, electricity and magnetism, and so forth. But Newton doesn't say anything about those in his first law; he just says that if an object accelerates, there has to be a force involved.

Newton recognized that moving objects near the Earth's surface experienced the force of friction, and so they slow down. That's why you seldom see anything approaching non-zero uniform motion in nature, because as soon as the ball starts rolling along the ground, friction causes it to slow down; that's the force that causes the deceleration, a negative acceleration, if you will. This idea that circular planetary orbits, as well as other accelerations are caused by an unbalanced force was radical in Newton's day, for it demanded that forces act over huge distances. After all, if a planet is going to orbit the Sun, if the Moon is going to orbit the Earth, then somehow forces have to be causing these accelerations, the change in direction. But Newton said, if you have an object moving around in a circular path, there clearly is a force acting on it because if I let go of this ball right now, a thing that would be sort of dangerous to do, it would fly off in a straight line. There has to be a force keeping it going in a circular path. That's the key idea related to Newton's first law of motion.

Now to Newton's second law of motion. This defines the exact mathematical relationship among three measurable quantities—force, mass, and acceleration. And he says, "The acceleration produced on a body by a force is proportional to the magnitude of the force and inversely proportional to the mass of the object." This law can be expressed as an equation: force equals mass times acceleration. Because mass is measured in kilograms (kg) and acceleration is measured in meters per second per second (m/s^2), that's in the modern system of units, the unit of force is a kilogram-meter per second squared ($kg\text{-}m/s^2$). This is called the "newton," that is, the force that is required to accelerate a one kilogram mass by one m/s^2. This isn't a very intuitive unit, the newton. What is a newton? So, to give you a better sense, this is the force exerted by gravity on an object about the size of a lime. That's the one thing in the grocery store I could find that seemed to be just about the right size. Another way of thinking about this, this may be an easier one, is that a newton is equal to the weight of about seven Fig Newtons. So, if you can remember seven Fig Newtons are equal to one newton, that's the force you have to exert to pick up those seven Fig Newtons. So, if that helps you remember it, fine.

Now, as you have more and more mass, you have to exert more force, more newtons. I'm probably exerting about 30 newtons right now, or maybe 50 newtons. A 22 pound mass would require 100 newtons to lift, and so forth. To counter the force of gravity that's pulling down on it. We'll talk more about gravity in the next lecture. That's one of the common forces around us.

This exact relationship—force equals mass times acceleration—enables us to calculate the force required, for example, to hit a 500 foot home run, or to lift the space shuttle into orbit. This is an incredibly useful equation: force equals mass times acceleration. Let's work a couple of examples just to show the power of this simple equation to solve everyday problems. I want to emphasize, I don't see mathematics as being the central role in this lecture series, but I think it's very important, right from the start, to recognize the power of mathematics in quantifying the natural world. We don't just talk qualitatively about forces and accelerations and masses. The scientific method allows us to derive exact equations, which then can be used to solve specific problems—build bridges, build buildings that don't fall down, that sort of thing—pretty useful to be able to use mathematics in conjunction with science.

So first, let's just calculate briefly the force it takes to accelerate a 1,000 kilogram sport utility vehicle to a speed of 30 m/s—that's about 65 miles per hour—in 10 seconds. So we have 1,000 kilograms. We want to get the 30 m/s in 10 seconds, so here's how we do it. The mass is just 1,000 kilograms; that's the mass of that vehicle you want to accelerate. The desired acceleration is 30 m/s in 10 seconds, and you can calculate the acceleration, m/s^2, by dividing 30 m/s by 10 seconds, that gives you 3 m/s^2. That's the acceleration. Now, if you apply Newton's second law—force equals mass times acceleration—force then is 1,000 kilograms times 3 m/s^2, that's 3,000 kg-m/s^2, 3,000 newtons. And you have to apply that force for 10 seconds to get up to the desired speed of 65 miles per hour. If you wanted to accelerate twice as fast, in 5 seconds rather than 10, you'd need to exert twice the force for 5 seconds—so, 6,000 newtons. That's why if you want a lot of acceleration you need to have a powerful engine that can exert a lot of force to power your car.

Let's try another example. Let's calculate the acceleration that's produced by a 6,000 newton force, a 6,000 newton rocket engine, on a 100 kilogram rocket. That's a pretty typical combination. Here you have to rearrange Newton's equation. We have acceleration now equals force divided by mass, and that's just flipping around the equation. So we have acceleration is equal to 6,000 newtons divided by 100 kilograms, and that gives us an acceleration of 60 m/s^2. That's a pretty rapid acceleration. That's the equivalent to an acceleration of 130 miles per hour every second. So, after 1 second the rocket is traveling 130 miles per hour, after 2 seconds it's going 260 miles per hour, up to 3 seconds, 390 miles per hour, and so forth. So faster and faster and faster, as long as the rocket engine keeps providing that thrust, that force.

Let's talk about mass for just a minute. Mass, in Newton's equation, represents an object's tendency to resist an acceleration. That's a property called its inertia. The more massive an object is, the greater the force that you need to apply to impose a given acceleration, and thus the greater its inertia. Now, if you want to get a sense of an object's mass, you don't have to have a gravitational force—they don't have to lift it against the force, though that's one way of doing it. If you're in deep space, you could take an object that's in your hand and just by trying to accelerate it, moving it back and forth or hefting it, what you're doing is you're feeling the inertia of the object, and you can do that in space just as easy as you can on the surface of the Earth.

And now we come to the third law of Isaac Newton. The most familiar but a very subtle law—"for every action there is an equal and opposite reaction." What Newton is saying is forces always act simultaneously in pairs. There are so many examples of this—forces acting in pairs. If you apply a force—throwing a ball—you exert the force on the ball, but the ball also exerts a force on your hand as you throw it. And these forces are equal and opposite, and they're simultaneous. If you crash your car into a tree, the tree exerts the exact same force on your car as the car exerts on the tree, and damage is done to both. You can't touch someone without having them touch you back is another way of thinking about it. You can see other examples of this, the simultaneity of forces acting in pairs. The famous example of Newton's cradle. You can apply a force by swinging one ball, and when you do this, instantaneously, the ball at the opposite end feels the force, and the forces go back and forth as the ball swings. We'll see this in other contexts later on, but you see forces acting in pairs. Even when I was swinging the ball around my head, here you see equal and opposite forces. The ball stays in that circular orbit. I'm exerting a force to hold the ball in place, but the ball is also exerting a force away from me, and these forces are equal and opposite. One is called centripetal acceleration, and the other is called centrifugal acceleration, and it's one way that Newton analyzed the motion of orbiting objects that was quite unique compared to other people in his time. Many examples, you've seen the recoil of a gun. Some of these examples are kind of subtle. The recoil of a rifle is rather subtle, but you're seeing lots of equal and opposite forces acting. You see the effects of an explosion that create a rapidly expanding cloud of gas. The gas pushes forward on the bullet, but the bullet pushes back on the gas. The gas pushes back on the gun, but the gun pushes back on the gas. The gun pushes your shoulder, but your shoulder pushes the gun, and these are all equal and opposite forces that are occurring more or less simultaneously.

One key idea: because the bullet is much less massive than the gun, it accelerates given that force—the equal and opposite force—to a much higher velocity. This is one of the reasons why, if you want to have a high caliber bullet, or a high velocity bullet, you have to have a fairly massive gun, because otherwise the recoil, if the gun was extremely lightweight, low mass, then the bullet would go in one direction, but the gun would really be kicked backwards too much the other way. So you have to have either a good should support, or a gun that's fairly massive.

In the liftoff of a rocket, you're also seeing consequences of equal and opposite forces. The rocket exhaust represents gases that are pushing on the rocket. But the rocket is also pushing on the gases, accelerating them in the opposite direction—you see the gouts of flames shooting out the backside of the rocket—they're accelerating just like the rocket is in the opposite direction. But in this case, the rocket is constantly being pushed by the expanding gases, the burning fuel, so it accelerates second by second, faster and faster and faster, whereas the gases aren't connected to each other, they are not physically tied and so they just sort of go off the other end. You have a slightly different case because one is solid, the rocket, and the gases that are expanding diffuse outward.

In this case, you can constantly accelerate a rocket just by putting a push on it, by constantly putting a force on it, and this strategy is very important in deep space exploration. A small steady force acting over very long times can accelerate a rocket to very high velocities, and thus you can, for example, use solar wind, or perhaps ion beams to produce the very modest acceleration, but they could be operating over years, and that would provide a long-term space voyage with very high velocities with the low acceleration acting over many, many years.

Now that you've seen examples of Newton's laws of motion, let me pose a question. I want you to think about this. Imagine yourself stranded at the middle of a perfectly frictionless sheet of ice. You're in the middle of this sheet of ice; you can't walk or crawl to move because there's no friction. How might you escape from that situation? What would you do? If you want to think about this, turn off the tape for just a second and ponder that question, how would you move off a perfectly frictionless piece of ice? Okay, for those of you who didn't bother to turn off the tape, or who are back with us, what you could do is remove a piece of clothing, like a shoe, and you'd throw it as hard as you could. You'd throw that shoe in one direction, and because of the equal and opposite force that that shoe would apply a force to you, and it would move you in the opposite direction. If the shoe was one-hundredth of your mass, whatever velocity the shoe had, you'd move in the opposite direction at one-hundredth of that velocity.

The equal and opposite forces of the third law leads to a new concept, that of momentum. The momentum of an object is defined as the product of its mass times its velocity, so momentum equals mass times velocity. It follows from Newton's laws of motion that, here's another quote, "in the absence of external forces, the total momentum of a system does not change." This is a key idea for Newton. Thus, when two objects collide, their velocities after the collision have to be arranged such that, the total momentum after the collision is exactly the same as it was before. Put in a mathematical form, you say that the sum of mass times velocity of the objects before is equal to the sum of the mass times the velocity of the object afterwards. And you've probably seen work examples in physics class of two pool balls of the same mass colliding and going off in different directions. That's all the consequence of the analysis of Newton's laws of motion. Newton's laws tell us that when two objects collide, even though each ball feels an unbalanced force, the entire system feels no force at all.

In physics the property is said to be conserved when it doesn't change. And I want to talk about this interesting property of conservation of momentum. It's a very important idea that certain physical quantities in the universe don't change for a closed system, and we're going to see this again and again and again as we look at different aspects of the physical universe—the conservation of angular momentum and momentum, conservation of mass, conservation of energy. These ideas provide a kind of symmetry, a kind of perfection to the universe, saying that there's certain quantities that do not change no matter what we in science can do. It gives a sense of an underlying order and beauty to the universe, these conservation laws.

A subtle and related idea that Newton studied is the law of the conservation of angular momentum—that is, an object that rotates will continue to rotate unless it is acted upon by a force. A spinning top is a great example. If you've ever seen a top, you can wind it up and it'll spin and spin, and unless friction slows it down at the very tip it'll just keep spinning. That's an example of the conservation of angular momentum. The Earth is a large spinning object. It rotates on its axis once every 24 hours, so the Earth has angular momentum. Over millions of years a very interesting thing has happened. The Earth has slowed down. Its rotation has slowed down. Early in its history, the Earth's day was only about 14 hours, 14 of our modern hours long. There were more than 600 shorter days per year early in Earth's history, but what's happened is the Moon's gravitation pull has slowed down the Earth, but in the process the Moon has gained angular momentum because the angular momentum of the Earth-Moon system has to be conserved. And so the Moon is going slightly faster, slightly faster, slightly faster, and so the orbit of the Moon has moved farther away from Earth as it has picked up angular momentum, and the Earth has slowed down. It is an amazing interactive system—all of which, though, can be analyzed by Newton's laws of motion.

To summarize, we've seen that Isaac Newton made major advances in understanding the motion of objects by recognizing two different kinds of motion. He had uniform motion—that's displayed by objects in which the velocity and the direction of movement don't change. And then you have acceleration, which includes any change in speed or in direction. Armed with these definitions, we come with the three laws of motion of Isaac Newton. First, nothing happens without a force; second, force equals mass times acceleration; and third, forces always act in pairs. Newton's three laws of motion together provide a complete framework for investigating and understanding all the forces of motions that occur in our lives. Newton, himself, applied this framework to understanding the ever-present force of gravity, and that's the subject of the next lecture.

Lecture Six
Universal Gravitation

Principle: *"Gravity is an attractive force that exists between any two masses, anywhere in the universe."*

Scope:

Isaac Newton's three laws of motion provide a firm mathematical foundation for identifying and studying the behavior of forces. During 1665–1666, Newton's remarkable years of discovery, he deduced the mathematical description of one of these forces—the universal force of gravity. Newton realized that an apple, attracted by gravity to the Earth, falls straight down. If thrown, the apple follows a parabolic path back to the surface in accord with Galileo. And, if launched with sufficient acceleration, the apple will adopt an elliptical Keplerian orbit, just like the moon.

Newton defined a mathematical description of this force: between any two objects there exists an attractive force of gravity that is proportional to the product of the two masses, divided by the distance between them squared. Newton's universal laws of motion and gravity revealed a deep, pervasive order to the natural world. One set of laws applies to Earth and the heavens, and the laws provide a framework by which many other phenomena can be studied.

Outline

I. Isaac Newton's three laws of motion define the concept of "force," a phenomenon that causes a mass to accelerate. Newton's laws establish the mathematical framework necessary for identifying and studying the behavior of any natural force.

   A. During 1665–1666, Newton's remarkable year of discovery, he deduced a mathematical description of the most pervasive of these forces—the universal force of gravity.

   B. To Newton's contemporaries, gravity was a terrestrial force, restricted to objects near the Earth's surface. Newton's great advance was to realize that one force—the universal force of gravity—acted on both an apple and the Moon.

- C. The apple, attracted to the Earth by gravity, falls straight down.
    1. If the apple is thrown, it follows a parabolic path back to the surface, in accord with the observations of Galileo.
    2. If the apple is launched with sufficient acceleration, it will adopt an elliptical orbit, as observed by Kepler for the Moon and planets.

II. Newton's investigations focused on finding a mathematical description of this force that would lead to the elliptical planetary orbits described by Kepler, as well as the law of falling bodies and the parabolic paths of cannonballs described by Galileo.
- A. Newton's universal law of gravitation defines the relationships among four measurable physical quantities: the mass of two objects, the distance between them, and the gravitational force thus generated:

$$F = G \times [m_1 \times m_2]/d^2$$

where G is the gravitational constant.
- B. Between any two objects there exists an attractive force of gravity that is proportional to the product of their masses, divided by the distance between them squared.

III. Newton's equation reveals that an equal gravitational force is experienced by any two objects—you and the Earth, for example, or the Earth and the Moon.
- A. Newton's equation for gravity has intuitive power.
    1. Gravity, in some deep and as yet mysterious way, is an attribute of mass.
    2. As the distance between two objects increases, the gravitational force between them drops off as one over the square of the distance—an "inverse square" relationship that is common in many everyday situations.
    3. This behavior suggests that we can imagine a gravitational field of lines radiating out from every mass.
- B. The numerical value of the gravitational constant, G, is the key to determining the Earth's mass and, by extension, the mass of other heavenly objects.
    1. Henry Cavendish (1731–1810), a chemist and physicist at Oxford University, devised an ingenious torsion balance to measure G in 1798.

2. He suspended a dumbbell with lead spheres from a wire such that the two suspended spheres were in proximity to two larger, fixed lead spheres. The slight force of gravity between the two pairs of spheres caused a measurable torsional force on the wire.
3. In 1998, an international group of scientists met to observe the bicentennial of Cavendish's discovery and to review new experiments devised to determine the value of G, now said to be $6.674 \times 10^{-11} \text{m}^3/\text{kg}/\text{sec}^2$.

IV. Newton's law of gravitation can be used to calculate many useful things.
   A. First, it is important to distinguish between weight and mass—two terms that are often interchanged.
      1. Weight is the gravitational force exerted by an object. Weight varies from place to place, depending on the strength of the gravitational field.
      2. Mass is the amount of stuff (measured in kilograms) of which an object is made; mass is invariant from place to place.
   B. Objects weigh less on the Moon because the mass and the radius for the Moon are different than for Earth.

V. Newton's universal laws of motion and gravity revealed a deep, pervasive order to the natural world. One set of laws applies both at the Earth's surface and in the heavens, and these laws provide a framework by which many other phenomena can be studied.
   A. Perhaps Newton's greatest legacy is a view of the universe as a place of deep mathematical order—a clockwork universe, whose mechanisms can be deduced through observation and analysis. This optimistic view was adopted in other human pursuits, including economics and politics during the Enlightenment a century later.
   B. Some followers of Newton, notably the French mathematician Pierre Simon Laplace (1749–1827), even speculated that, since the laws of motion are exact and each particle in the universe has a measurable position and velocity, the future is preordained. Some scholars even called into question the nature of free will.

Essential Reading:

Trefil and Hazen, *The Sciences: An Integrated Approach*, Chapter 2.

Supplemental Reading:

Andrade, *Sir Isaac Newton*, Chapters 3–5.

Cohen, *The Birth of a New Physics*, Chapter 7.

Questions to Consider:

1. If gravity is always an attractive force between any two objects, why does a helium balloon fall up? (Hint: Think about Newton's laws of motion.)
2. When a space ship accelerates in the vacuum of space, what does it push against?

Lecture Six—Transcript
Universal Gravitation

Isaac Newton's laws of motion provided a simple and a unifying framework for analyzing motions on Earth and in the heavens. Newton made a conceptual breakthrough by recognizing two different kinds of motion—uniform and accelerating. His laws of motion did something else as well. They defined a force as any phenomenon that causes an object to accelerate. Newton applied this definition to his own studies of gravity, which is an attractive force that exists between any two masses anywhere in the universe.

In Lecture Six, I want to tell you about the universal force of gravity. First, we're going to see how Newton's ideas came to define and helped him recognize the force when he saw it. Then I'm going to introduce the mathematical expression for the gravitational force, and I'll try to give you some sense of the intuitive appeal and the quantitative power of this expression for gravity. Finally, I'm going to explore some of the ways that Newton's clockwork vision of the universe has influenced our own perception of our place in the cosmos.

Newton's analysis of the force of gravity was rooted in his understanding of the relationship between motion and force. Let's begin by reviewing Newton's three laws of motion, remember the laws we discussed in the last lecture. Recall that Newton broke with tradition when he defined two different kinds of motion. First of all, uniform motion, which is an object moving at a constant velocity in a constant direction, or an object at rest sitting on a table, for example. And then, the other kind of motion, acceleration, which is any change in either the speed of an object or in the direction of its movement. So unlike previous thinkers, Newton said that orbital motion, perfectly circular motion at a constant speed is acceleration, not as previous thinkers had said, uniform motion.

And then we have the three laws of motion. The first law states that nothing happens without a force. An object remains in uniform motion unless it's acted upon by a force, and if it's acted on by an unbalanced force, then it will accelerate. The second law puts this idea in quantitative terms, it says force equals mass times acceleration, and you can plug numbers into that equation and come up with useful conclusions. We saw the power of that equation in the last lecture. And then third, the third law presents the idea that forces act in pairs. Equal and opposite forces occur simultaneously. When you push on an object, it pushes back on you with the same force at the same time.

As powerful and as original as these ideas were, they are not in and of themselves the end point in the study of motion. In fact, they're really just the beginning. By defining the notion of force, Newton's laws marked the starting point for the research that continues to this very day in forces and motions in the universe. See, Isaac Newton's three laws define this concept of force—a phenomenon that causes a mass to accelerate. Newton's laws establish mathematically the framework necessary for identifying all kinds of forces in our everyday life. You just have to look around you. You observe objects as they change what they're doing; they do interesting things. They accelerate. When objects do interesting things, there has to be a force involved. And so you can go on and look at nature and say, where do I see objects accelerating? And then what's the force involved in that acceleration?

During those key years of 1665 and 1666, which saw Newton's remarkable burst of discovery, he deduced a mathematical description of one of these forces—the most pervasive force, the universal force of gravity. Recall that during those years, 1665 and 1666, the bubonic plague had struck England, and that's the period when Newton retreated to his family farm as a consequence of Cambridge University being shut down for those years. And on the farm he had a year and a half to think and reflect, to ponder the things he'd learned about Kepler's laws, about Galileo's ideas, about other concepts that he had studied as an undergraduate at Cambridge, and he had a year and a half just to think about them.

Let's talk about gravity. To Newton's contemporaries, gravity was a terrestrial force; it was restricted to objects near the Earth's surface. But Newton realized that gravity is a universal force. It extends all the way out to the planets, to the Moon, to the stars. He realized this while studying in the family apple orchard. The young scholar looked up to see an apple ripening on the tree, and above it he saw the Moon in its orbit. Newton's great advance was realizing that it's a single force—one force that's acting on both of these objects.

In later years, Newton told friends that this discovery came to him "as he sat in a contemplative mood," and that it was "occasioned by the fall of an apple." Subsequent descriptions told about the apple hitting Newton on the head—and I don't think that actually happened—but you can imagine the young Isaac sitting there thinking about the universe and how it worked, seeing the apple, and seeing the Moon, and the apple falls. He says, "Why doesn't the Moon fall? It's up there also. It's just a little farther away." Here's the gist of Newton's idea. You have an apple. It's in the tree. And then suddenly that apple breaks loose and it falls to Earth, straight down. Well, you could also pick up that apple, and you could throw it sideways with a certain amount of horizontal velocity, and if you do, as Galileo tells us, the apple adopts a parabolic path. So, when you throw it, it has a curving path. As you throw it harder and harder and harder, the apple adopts more horizontal distance and goes farther and farther. What Newton realized, if you threw the apple hard enough, the apple would go into orbit. It wouldn't stop falling, it would continuously fall, but as it fell it would move horizontally, so it's falling, but it moves horizontally and it keeps going around the Earth. That's what the Moon's doing, it's going around the Earth, constantly falling, but it has sufficient horizontal velocity to keep it in orbit.

Previous scholars had seen these same events. These were familiar to everyone, but they saw them differently. Newton saw them as variations on a single theme. The Moon is falling. The apple is falling. Objects that you throw are falling. Only when you add significant horizontal or sideways velocity can you reach that orbital situation of the Moon. So it is with any planet, with any moon, anything that's orbiting the Sun, anything that's orbiting the Earth is falling.

This idea was vividly brought home to me with the pulp science fiction story. When I was a young boy, I was probably nine or ten years old when I read this story. And I want to tell you the story. As I remember it, the action took place on an asteroid someplace out near Mars or Jupiter, and a few Marines, three Marines, were defending Earth against aliens. The aliens had landed on this asteroid, and I believe there were two of them and they had their spaceship on this asteroid, and they had discovered Earth and they were about to go back to their home planet and launch an invasion against Earth. So these three Marines were the last bastions of hope for the survival of planet Earth. Two of the Marines were big hulking fellows, and they were lobbing their dwindling supply of grenades at the aliens who were crouched behind rocks on the asteroid. The third Marine was a scrawny fellow. He had glasses, and he was sitting down. He had a calculator, or probably a slide rule in those days, and he was calculating, fiercely scribbling numbers down on a piece of paper. And as the supply of grenades dwindled down to the last two or three, he reached down, he picked up a grenade, he turned the opposite way, and he threw it in the wrong direction. His two colleagues looked as if, "What are you doing? That was one of our very last grenades." And the young Marine said, "Just wait." And you guessed it. The grenade went around the opposite side of the asteroid where there weren't any mountains or hills in the way, landed on the space aliens, exploded, killed them, and the Earth was saved by this little piece of Newtonian mechanics. By the way, I can't for the life of me figure out where this story was published, or who the author was. If any of you watching these tapes know, please get in touch with The Teaching Company and let me know. I'd love to find that story again; it was a very vivid childhood memory.

Well, Newton's concept of gravity is simple; it's compelling; but the true elegance lies in the mathematical equation, which Newton derived to explain this force. It wasn't just a qualitative description of the cosmos, it was quantitative. He described a force in terms of four measurable quantities, four variables. The first one is the mass of an object. The second variable is the mass of a second object. The third variable is the distance between these two objects. And finally, we have the force—the gravitational force that results. And this is the equation that Newton came up with. He said: force equals a constant—a capital G for the gravitational constant—times the first mass, times the second mass, divided by the distance squared ($F = G \times [m_1 \times m_2]/d^2$). That's Newton's equation for the universal force of gravity. Stated in plain English: Between any two objects there exists an attractive force of gravity that's proportional to the product of their masses, divided by the distance between them squared.

Now, Newton used rather complicated mathematical reasoning, and he demonstrated that stable orbits are possible only if you have a 1 over d^2 kind of relationship. If you have an exponent less than 2, it leads to a steadily decaying orbit because the force doesn't drop off sufficiently with increasing distance. And if you have an exponent greater than 2, 2.1 or 2.2, for example, it allows the orbiting body to escape because the force drops off too quickly and the body just keeps moving outwards. Only with 1 over d^2 do you get the exact relationship.

Newton's powerful equation reveals that an equal gravitational force is experienced by any two objects, you and the Earth, for example, or the Earth and the Moon. The exact same gravitational force is exerted on both bodies. In fact, when an apple falls to the Earth, the Earth also falls a miniscule distance towards the apple. And there's a kind of lever law here. If two objects have exactly the same mass, they would fall the same distance towards each other, but if one is more massive, it's sort of like a situation on a seesaw when you have two children that don't weigh the same amount, that the heavier one has to sit closer to that fulcrum point and the one who's farther away, experiences a much larger motion on the seesaw. It's the same sort of thing. So Newton saw that lever arm kind of effect in gravity.

Several aspects of Newton's law for gravity have incredible intuitive power. First, gravity in some sort of deep and mysterious way is an attribute of mass. More mass means more gravitational force, and it's in direct proportion. The gravitational force thus bears a relation to mass, just for an example, as your buying power does to the amount of dollars that you have in your pocket. There's this direct relationship between force and mass. Pick up an object and you feel the force that it exerts against your hand. And think about mass from this perspective. In a sense, we can describe mass, therefore, in terms of its phenomenology, but we don't really understand what mass is at a deeper level. We can just describe this phenomenon.

Now let's look at the bottom half of Newton's equation, that's the 1 over d^2 part of it. As the distance of two objects becomes greater, the gravitational force between them drops off. And it doesn't drop off just in proportion to distance, but it drops off in proportion to the square of the distance. That's called an inverse square relationship and is actually a very common relationship. You see this many, many times even though you may not recognize it. In a sense, distance dilutes gravity in the exact same way that distance dilutes the light from a flashlight. Let me show you this situation. If you have a flashlight shining on a surface and it puts out a certain amount of light. Now, if I doubled the distance, the area of that light, the bright spot, the diameter increases by a factor of two, but the area increases by a factor of four. I doubled the distance, I have four times the area, which means the same amount of light is spread out over four times as much area. If you had a light meter, the difference between the intensity at one distance, and a distance twice as far away would be the ratio of one to one-quarter, or 1 over d^2. It's the exact same thing with gravity.

It suggests that you can imagine a gravitational field, an imaginary array of lines that radiates straight out from any object, and essentially the gravitational force is the number of those lines that intersect between one mass and another. As you double the distance the number of intersections of lines becomes less. So you can think about gravity in this geometrical way.

Even after Newton proposed his equation for gravity, there was one big gap, and that was the numerical value of this capital G, this gravitational constant. This constant is tremendously important. It helps you estimate the magnitude of the gravitational force between any two objects. It also allows you to calculate things like the mass of the Earth and other heavenly objects. And we'll see how that's done in just a second. Determining G is extremely difficult. It's difficult because your experiment has to be completely independent of the Earth's huge gravitational force that swamps out most other measurements. The experiment also has to eliminate any other contributions by stray electrical or magnetic fields, and that's not very easy. After all, a slight static electric charge on a comb can pick up a piece of paper. You can have a magnet, which is a very tiny device compared to the entire Earth, and yet against the Earth's entire gravitational force, the magnet picks up objects. It counters the entire pull of the Earth. So if you think about gravity, it's a very weak force compared to some of the other forces that are around us all the time.

The most famous attempt to determine G was undertaken two centuries ago by the English chemist Henry Cavendish. He was born in 1731 in Nice. He was born into one of the wealthiest families in all of Britain. He was educated at Cambridge University and spent most of his life, more than 50 years, in London performing scientific experiments. Cavendish is said to have been uncommonly shy, especially shy of women, and he never married. There is a story that he once inadvertently met a housemaid on the stairs of his very large home. He was so unnerved by this experience, that he had a separate private stairway built, just to avoid future incidents of that sort. He died a recluse in a London home in 1810.

Of his many important experiments, Henry Cavendish is best remembered for devising a very ingenious method for determining the gravitational constant G. He did this in 1798. He suspended a dumb bell with lead spheres from a wire, such that the two suspended spheres were in proximity to two much larger lead spheres that were fixed. The slight force of gravity between the large spheres and the smaller lead spheres caused the suspended dumb bell to rotate, to torque slightly. Well, Cavendish knew how much torque it took to twist his wire. He measured that twist, and therefore he was able to measure the force of gravity because he knew the two masses of the lead weights. He knew the distance between the weights, and he measured that twisting force. So he could solve for G. Substituting that measured value for G, the gravitational acceleration of the Earth's surface and the Earth's radius all into a modified form of Newton's equation yields the Earth's mass.

And I want to do this calculation for you because this is really interesting. You can actually calculate the Earth's mass, just by knowing the G and a few other simple things. Let me show you one way of doing this calculation. You have to think about the force exerted by the Earth on a mass, any mass will do. As you think about the force—oh, from Newton's law for gravity, force equals this mass times the mass of the Earth divided by the radius of the Earth squared, and also with that G as a constant. But from Newton's second law you can do this another way. You can also say that force is equal to the mass times the acceleration due to gravity, because force equals mass times acceleration, and the acceleration due to gravity at the Earth's surface is 9.8 m/s^2, that's g, or the acceleration due to gravity. So you have two equivalent descriptions of force—one using Newton's equation for gravity, and one using Newton's second law: force equals mass times acceleration. Because these two equations describe the same force, you can set them equal to each. So you say, mass times gravity, the little g, is equal to mass, the same mass, times capital G times the mass of the Earth divided by r^2. The first thing you can do is cancel the mass, the little mass, on both sides—just cancel them out. So now we have the gravity g—small g, which is the acceleration due to gravity—is equal to capital G, which we know, times the mass of the Earth divided by the radius of the Earth squared, and we know the radius of the Earth. So, you can solve this equation by rearranging it. Solve for the mass of the Earth. Indeed, the mass of the Earth is equal to little g times r squared over big G [$(g)(r^2/G)$]. Plugging in the numbers, you end up with the mass of the Earth 6 times 10^{24} kilograms. The Earth weighs six trillion, trillion kilograms. That number comes just from Cavendish's discovery of the value of the big G, the gravitational constant.

Cavendish presented this calculation for the first time in his now classic paper, *Experiments to Determine the Density of the Earth*, published in 1798. That paper stands as a great landmark in geophysical science. It might surprise you to learn that the measurement of this capital G, the gravitational constant, is still of great interest to scientists around the world. In November 1998, a group of 45 physicists met in London to celebrate the 200th anniversary of Cavendish's original experiment, and they also compared notes on several new measurements of the constant. All these workers performed meticulous experiments. They enclosed their apparatus in vacuums. They eliminated all stray magnetic fields and electrostatic fields. They checked and rechecked every aspect of their experimental work. Rival groups in France, Russia, and the United States used modern versions of the Cavendish experiment—that is, masses suspended either on a wire or on a strip of metal, and watched how that wire or strip of metal twisted due to gravitational force.

But there were a couple of new ways of doing it as well. There was a group in Zurich that weighed a kilogram mass on an exquisitely sensitive balance. In one case, they did it with a 1,000 kilogram mass just below sort of a donut-shaped mass, and then they raised that donut-shaped mass and put it just above the kilogram mass, and they looked at the difference in the weight of the object when that 1,000 kilograms was just below, or just above the suspended mass on the balance. A very difficult measurement, but they were able to do it and come up with a value for G.

Another American group tried a similar approach. They measured the time that it took an object to fall, and you can measure times to millionths or billionths of a second. So if you measure the time it takes an object to fall with a heavy mass below, as opposed to a heavy mass above, you get a slightly different time. Again, you can solve for that capital G, that gravitational constant.

All of these techniques now yield similar values for G. For the record, the best estimate of G is now about 6.674 times 10^{-11}

$m^3/kg/sec^2$. So it's a number that people are still converging on, but that's pretty much the accepted value now.

Newton's law of gravitation can be used to calculate many useful things. First of all, given the universal nature of the gravitational force, we have to recognize the fundamental difference between weight and mass, and I do want to make this distinction. These terms are often interchanged, weight and mass. Weight is the gravitational force. Weight is the force of gravity exerted on your hand when you pick up something. Weight varies from place to place. It varies from place to place on the Earth's surface. It's different on the Earth. It's different on the Moon. Mass, however, is the amount of stuff, the number of atoms, and mass doesn't vary from place to place. The mass of this object is constant even though the weight may vary depending on where you are.

You can calculate the weight of a 60 kilogram person on Earth, for example. And we just used the expression, force equals mass times acceleration. A 60 kilogram person times the acceleration at the Earth's surface, 9.8 m/s^2, gives you 588 newtons. But you can do the exact same calculation by plugging into the expression for gravity, force equals capital G, the mass of the Earth, times the mass of the object you're looking at divided by r^2. And plugging in those numbers, you come up with the exact same answer, 588 newtons. Two equivalent ways of calculating the force, the weight of an object.

But objects weigh less on the Moon, because the Moon is a less massive object than the Earth. And so, you can do the exact same calculations and try to determine what is the weight of an object at the Moon's surface. Let's do the same thing. For a 60 kilogram individual, what would their weight be on the surface of the Moon? And here you have to know the radius of the Moon, and you need to know the mass of the Moon as well. Those are both two numbers that we have down, so you use the same equation for gravity: force equals mass of the Moon times the mass of the individual divided by the radius of the Moon squared. Of course, the gravitational constant is in there, you could plug in numbers again, and you find out that on the Moon, a 60 kilogram person only weighs 95 newtons. Ninety-five newtons is only one-sixth of the weight of an object at the Earth's surface. You may remember those wonderful pictures of humans walking on the Moon and some of the antics that they displayed, skipping and jumping, and it looked like everything was taking place in slow motion. That's because the force of gravity was so much less, their weight was so much less, that they were able to jump higher. They didn't accelerate nearly as fast, only one-sixth the acceleration, so they'd jump up and they'd come down much more slowly. You could hit a golf ball a lot farther on the Moon, largely because of the difference in the gravitational acceleration at the surface of the Moon, also because there's no atmosphere, so there's no wind resistance to that golf ball. You'd have to have a very large golf course to have a par 5 hole on the Moon. Maybe some day they'll do it. I can imagine lunar golf would be a lot of fun.

I want to close this consideration of Newton's laws on a more philosophical note. Newton's universal laws of motion and gravity reveal the deep and a pervasive order in the natural world. One set of laws applies both at the surface of the Earth and in heavens, on the Moon and on the planets and stars. These universal laws occur everywhere. They provide a framework by which you can understand many other phenomena, and they provide a unity to the understanding of the natural world. There's nothing special about the surface of the Earth in terms of the laws of motion, in terms of the forces involved. Perhaps Newton's greatest legacy is the view of the universal as a place of deep mathematical order, a clockwork universe. The mechanisms can be deduced through observation and through analysis of the natural world. In the words of the distinguished historian of science, I. Bernard Cohen, "The greatest achievement of Newtonian science must ever be the first full explanation of the universe on mechanical principles. One set of axioms and a law of universal gravitation that applied to matter everywhere, on Earth as in the heavens. Who, after studying Newton's magnificent contribution to thought, could deny that pure science exemplifies the creative accomplishment of the human spirit at its pinnacle." What an exalted view of science and what a transforming view Newton gave to all humans in all different endeavors.

The optimistic view that humans could deduce the order of the natural world had a significant trickle down affect in other human endeavors. During the enlightenment, which was about a century after Newton, scholars advocated this rationalist approach to all different kinds of human endeavors—to economics, to education, to political systems, to the law. It was felt that there must be universal laws that dictated all kinds of activities that humans undertook. Indeed, Newton's first firm statement of this cause and effect relationship in the natural world seems to have pervaded our legal system today. If something bad happens, there must be a specific cause. Whatever bad happens, you can point to a cause, and indeed you can go out and sue someone, which is what it seems to be these days. But indeed, cause and effect has become part of our way of thinking about the universe. The drafters of the United States Constitution and the Bill of Rights firmly believed that there was a rational system of government that could be derived, that there were laws and ordering to the political system as well. The Bill of Rights and the Constitution thus reflect this optimistic view of order—the attempt that humans could find that order, and in fact run their lives according to those laws.

Newton's laws even contributed to theological debates. Some followers of Newton, notably the French mathematician Pierre Simon Laplace, who lived from 1749 to 1827, and about whom we're going to hear a lot more in Lecture Thirty-Four, he even speculated that since the laws of motion are exact, and they predict the motions of every possible particle under every kind of force and interaction, that if you knew what every particle was in terms of its position, in terms of its momentum, its velocity, of the force acting on it, that indeed the future of every particle in the universe was preordained, and this lead to a very interesting theological speculation. Is it possible, since everything in the universe is preordained, that there is no such thing as free will; that we act out our lives; that every event in our lives—my standing before you today and delivering this lecture—was something that was preordained just by the way particles were set into motion billions of years ago before anyone had every heard of Isaac Newton and his laws of motion? This idea, by the way, has been silenced by more recent findings of inherent uncertainties in the natural world.

Let me summarize this lecture on the universal law of gravity. We've seen how Newton's laws of motion provided a framework for identifying forces of nature, and then Newton went on himself to focus on this force of gravity that you could see everywhere. Gravity is the force that operates between any two objects anywhere in the universe. The magnitude of the force is proportional to the mass of the two objects, it's inversely proportional to the square of the distance between those two objects, and the universal gravitational constant G provides the appropriate scaling factor.

In the next lecture we're going to see how Newton's legacy extends to the concept of energy.

Lecture Seven
The Nature of Energy

Principle: *"Energy is the ability to do work."*

Scope:

Newton's laws not only provided a systematic procedure to study forces and motions, but they also instilled in the scientific world an unswerving confidence in nature's deep order. Yet Newton and his contemporaries were unable to solve the riddle of heat, and its companion phenomenon light, which baffled researchers until well into the 19th century.

The definition of energy depends on the familiar concept of work, which is the exertion of a force over a distance. Energy is the ability to do work—the property of a physical system that allows it to exert a force over a distance. Energy comes in a wide variety of forms. Objects in motion possess kinetic energy. Many systems store potential energy, which is poised to do work. Waves possess energy that can be transferred without significant movement of the medium.

Benjamin Thompson (1752–1814) demonstrated that heat is a form of mechanical work and thus is equivalent to energy. Radiation, such as light from the Sun, is another energy form, which can travel through a vacuum at 186,000 miles per second. Finally, one of the defining discoveries of modern science is the equivalence of mass itself with energy—the energy released in nuclear reactions.

Outline

I. Newton's laws of motion define a systematic procedure to study forces and motions, but Newton was not able to solve the riddle of heat, and its companion phenomenon, light, which baffled researchers until well into the 19th century.

   A. The study of energy is called thermodynamics. Unlike Newton's laws of motion, the laws of thermodynamics did not spring fully formed from one mind. Rather, these ideas emerged gradually from the work of many researchers, and it was only later that they were set down in the form we now know.

B. Consider two age-old questions: What is heat? And what is light? Even though heat and light are among the most tangible of all physical phenomena, their origins are among the most subtle and difficult to deduce. Their understanding arose from the abstract concept of energy.

C. The definition of energy requires the new concept, work, which has a very specific meaning in physics.
   1. Work is expended when a force causes a body to move through a distance: work equals force times distance.
   2. The unit of work is the joule, defined as a force of one newton exerted over a distance of one meter (a newton-meter). In the English system, the associated unit of work is the foot-pound.

D. Energy is the ability to do work—the ability to exert a force over a distance. Energy is a measurable attribute of physical systems, but, unlike mass or motion, it has an intangible quality.

II. Energy was difficult to recognize and describe, in part because it can adopt so many different forms. There are many ways to exert a force over a distance. Even so, three forms of energy—kinetic, potential, and waves—have long been recognized for their ability to do work.

A. The most obvious form of energy is carried by objects in motion—kinetic energy.
   1. The kinetic energy of any object can be calculated from its mass and velocity: force equals one-half of the mass times the velocity squared or:

   $$F = \frac{mv^2}{2}$$

   2. This equation allows us to calculate the relative energy of a car traveling at different speeds.

B. Many natural systems store energy, called potential energy.
   1. The most obvious form of potential energy is due to gravity—the stored energy of water behind a dam or the weight of a grandfather clock.
   2. Other common forms of potential energy are chemical energy (in gasoline and food), electrical potential (in batteries and your wall outlets), magnetic potential (in a

refrigerator magnet), and elastic potential (in a tightly stretched rubber band).
- C. Waves represent an efficient form of kinetic energy, in which energy can be moved long distances with only small movements of mass.
  1. Sound waves transmit energy in the air.
  2. Waves at the surface of the ocean can carry energy across thousands of miles.
  3. Earthquakes can transfer immense amounts of energy and cause devastation to inhabited areas.

III. It took much longer to discover three additional, far more subtle forms of energy: heat, light, and mass.
- A. The nature of heat was a matter of intense debate for centuries. Many noted scientists, including the influential French chemist Antoine Lavoisier (1743–1794), favored the caloric theory, which described heat as a massless fluid that could flow from object to object.
- B. The debate was ultimately resolved by the opportunistic American-born inventor, Benjamin Thompson (1752–1814), known as Count Rumford.
  1. Rumford had always been fascinated by the phenomenon of heat, and he conducted many studies to improve its use in everyday life. His invention of the modern style cookstove and fireplace gained him fame and a degree of wealth.
  2. Rumford was appointed commandant of police in Bavaria, where one of his duties was to oversee the manufacture of cannons. Rumford noted the unremarkable fact that boring cannon produced heat. Rumford realized that the heat was produced by friction—mechanical action.
- C. The caloric theory was buried forever when British scientist Humphry Davy, a brilliant public lecturer, succeeded in melting ice just by rubbing two pieces together on a cold winter day.
- D. Another form of energy is radiation—light traveling 186,000 miles per second. The principal source of energy at the Earth's surface is radiation from the Sun, which travels across the vacuum of space.

E. Finally, one of the defining discoveries of modern science was Albert Einstein's recognition in 1905 that mass is also a form of energy, according to the familiar equation, $E = mc^2$, where c is the speed of light.

Essential Reading:

Trefil and Hazen, *The Sciences: An Integrated Approach*, Chapter 3.

Supplemental Reading:

von Baeyer, *Maxwell's Demon*, Chapters 1–2.

Questions to Consider:

1. Work and energy are both measured in the unit joules—a force times a distance. Does that mean that work and energy are the same thing? How are these two physical concepts related?
2. Is it possible that there are as yet undiscovered forms of energy? What might be the implications of such a new form?

Lecture Seven—Transcript
The Nature of Energy

Newton's laws provided the scientific world with a remarkable ability to study forces and motions in nature as well as an unswerving confidence that various other laws could be deduced by observing that natural world. But Newton was not able to solve the riddle of heat and its companion phenomenon—light, which baffled researchers well into the 19th century. This lecture is going to explore the nature of energy, which is defined as the ability to do work.

My two principal objectives in this lecture are to first define the subtle concepts of work and energy, and then to catalogue the many different kinds, these surprising and interchangeable kinds of energy. These forms of energy include kinetic energy, which is the energy of motion; potential energy, which is stored energy—energy waiting to be used. We have wave energy, including waves in solids and liquids and in gases. There's heat energy, the energy of atomic motions. And then the more subtle forms of energy, light energy, which travels at 186,000 miles per second. And finally, the energy associated with mass.

What attributes of nature enable a system to change? What allows you to exert a force? These are deep and profound questions, and their answer lies in the study of energy—a scientific field that's called thermodynamics. Now, that word thermodynamics comes from two roots: thermo for heat, and dynamics for motion. So thermodynamics is quite literally the study of heat in motion, how heat moves.

Unlike Newton's laws of motion, the laws of thermodynamics did not spring forth fully formed from one mind. There was no eureka moment, as it were. Rather, these ideas emerged gradually from many, many different researchers. It was primarily in the early- and mid-19th century that the laws of thermodynamics were finally formulated and understood in a systematic way. It was actually only much later that they were set down in a form that we use today, in a form that you and I can understand in just everyday language, and that's what I'm going to try to do in the next few lectures.

The difficulty in understanding energy is epitomized by a simple candle. Light a candle. Look at the flame and ponder two questions, age-old questions. What is heat? What is light? Are heat and light a kind of force? Can they cause an acceleration? Well, heat is certainly associated with some kinds of motion, but hot objects don't always accelerate. Are heat and light in themselves a kind of motion? If so, what forces cause the acceleration of heat and light? Do heat and light have mass? Do they have some kind of substance, physical substance? Even though heat and light are amongst the most tangible objects and phenomena in our material world, their origins are amongst the most subtle and difficult to deduce.

Ultimately, an understanding of both heat and light depends on the essential, but very abstract concept called energy. The definition of energy requires introduction of another concept, and that concept is work. It has a very specific meaning in physics. Work is one of those everyday words that we run into all the time in science, which actually has a very specific scientific meaning, and a more colloquial meaning in our everyday life. We've seen several of these terms. Motion is one, force, acceleration, and energy—those are all terms that we use in everyday life, but they have very specific scientific meanings. Work is expended whenever a force acts over a distance, and the formal definition of work is a force times the distance, a force acting over a distance.

The unit of work is a joule, and that's when one newton is exerted over the distance of one meter—a newton-meter, if you will. In the English system, the appropriate unit for work is the foot-pound, but I want to refer to newtons and joules. So, if you have a lime, which weighs—remember, the force that it exerts is about one newton—and you raise it one meter, you've done a joule of work, or alternatively, if you'd like to think in terms of Fig Newtons, seven Fig Newtons make one newton, and one meter—we've done a joule of work.

Work is only done if the force is exerted in the same direction as the object's motion. So the Earth does no work on the Moon because the Moon's movement is always at right angles to the gravitational force. So if you have a swinging ball—the kind swinging over your head—once it gets going, once it starts that circular motion, and I stop, it'll keep going forever unless there are frictional forces and other forces that slow it down. I'm not doing any work at all now, because the ball's moving at right angles to the direction of the force.

You do work if you push your car along the road. If your car runs out of gas, as long as you're pushing it and it's moving, you're doing work. But, if the car doesn't move, no matter how hard you push and no matter how much sweat you work up, you're not doing any work. By the same token, if you lift a heavy weight, you're doing work. But, if you're just holding it out, no matter how tired your arm gets, since you're not exerting a force over a distance, you're not doing any work. That's the formal definition of work.

Now we come to the definition of energy. Energy is defined as the ability to do work. That is, energy is defined as the ability to exert a force over a distance. So, any physical system, any phenomenon that allows you to exert a force over a distance is a kind of energy. Energy is the measurable attribute of physical systems, but unlike mass or motion, it has a mysterious intangible quantity, you can't really hold energy, per se, in your hand, but you can hold objects that have the potential to exert a force over distance, so those objects have the attribute of energy, they contain energy in some sort of way. It's little wonder that this concept of energy was so poorly defined until the middle of the 19th century that it took scientists so long to come to terms with it. Energy was difficult to recognize and describe in part because it can adopt so many different forms in our surroundings. There are many, many different ways to exert a force over distance. Let's think about the kinds of motions that we see around us and analyze what form of energy then is involved in those motions. There are three really common kinds of energy in our everyday life. They are real obvious, and these are the ones that scientists first recognized. These are called kinetic energy, potential energy, and wave energy. These have long been recognized for their ability to exert a force over a distance.

Perhaps the most intuitively obvious form of energy is that carried by objects in motion, kinetic energy. If you flick a crumb off your sleeve, or throw a ball, the ball goes through a window, it can break things, it can exert a force over a distance. It can do work, and therefore the ball in motion, your finger in motion, anytime you move something you're causing kinetic energy, which has the ability to do work.

The kinetic energy of any object can be calculated from its mass and its velocity, and it turns out, in this case, force equals one-half of the mass times the velocity squared [$F=(½)(mv^2)$]. This formula is easily derived from Newton's second law, which says force equals mass times acceleration [$F=m×a$]. And you need a couple other definitions, classic definitions as well. You need to have the standard formula for acceleration, velocity, and distance that relates to these three parameters. For example, distance equals one-half acceleration times time squared, $[d=(½)(a×t^2)]$, and also velocity equals acceleration times time, $[v=a×t]$. And with those three standard definitions, you can derive this idea of kinetic energy.

Here's how you do it. Kinetic energy has to be force times distance, $[KE=F×d]$. Well, the force is just simply mass times acceleration, $[F=m×a]$. And distance, d, is the distance traveled. But distance, from our previous equation, is equal to $(½)at^2$, $[d = (½)(a×t^2)]$. So that means that kinetic energy equals $(½)m(at)^2$, $[KE=(½)m(at)^2]$. Now this one other trick, remember the velocity is equal to a times t, $[v=a×t]$. So here we substitute (a×t) with velocity, and we get kinetic energy equals $(½)mv^2$, and that's the equation for kinetic energy.

This equation allows us to calculate the relative energy of a car traveling at different speeds, for example, and you may be surprised at the result. Think, for example, about the kinetic energy of your car when you travel 35 miles per hour, compared to 55 miles per hour, compared to 75 miles per hour. In the equation for kinetic energy, the one-half M part cancels for all these three, so all you have to worry about here is just the relative amount of energy is proportional to v^2. So let's look at this, 35×35 is 1,225. If you go up to 55 miles per hour, 55×55 is 3,025, more than twice the kinetic energy, and you go up to 75 miles per hour, 75×75 is 5,625. That's more than five times the kinetic energy than when you're traveling at 35 miles per hour. So, you may just approximately double your speed, but you have many times more kinetic energy. Because kinetic energy is proportional to velocity squared, that means as you increase in slight increments your velocity, you can greatly increase your energy. And that has consequences.

There are two in particular. One, as you go faster and faster, you're going to consume more gasoline because you have to put the energy into the car to get that kinetic energy. See, your gas consumption, your miles per gallon, is going to start dropping off dramatically as you go to very high speeds.

The second thing is the danger if you get into an accident, because when you get into an accident, all your kinetic energy gets converted into other kinds of energy, the smashing of your car. At higher velocities there's much more energy that has to be converted into the damage of the car, and therefore, high-speed crashes are proportionally much more damaging than low speed crashes. It's not just a simple linear proportion to velocity, it's velocity squared.

Okay, let's look at a different kind of energy. Many natural systems store energy, and that energy is waiting to be used, just waiting to exert a force over a distance. These varied forms are called potential energy. The most obvious form of potential energy in everyday experience is due to gravity—the stored energy of water behind a dam, for example, or the weight of a pendulum, of a weight in a grandfather's clock that drives the grandfather's clock, that's stored energy, gravitational potential energy. And it's energy that's there by virtue of a mass's position within the Earth's gravitational field. If its up high, its waiting to drop and release energy, converting gravitational potential energy, for example, into kinetic energy.

To raise the mass, m, to a height, h, takes work, and work is force times the distance, so you can calculate gravitational potential energy just by our simple definition. Energy equals work, which is force times distance. But the force of gravity on an object is just mass times g, so that's mass times little g, gravitational acceleration, times height. So you can use this equation, $(m \times g) \times h$, as the energy contained by an object to make all sorts of calculations. For example, a two kilogram mass, if you want to lift that one meter, how much energy does it take? Well, you just substitute into the equation. Two kilograms times 9.8 m/s^2, that's the acceleration due to gravity, times one meter, is 19.6 joules. So that's the energy required to raise a two kilogram mass one meter.

There are many other kinds of potential energy that are very common in our day-to-day life. We have chemical energy stored in food—chemical potential energy—or in matches that you'd use to light a candle. We have chemical potential energy also in batteries. There are chemicals here ready to release energy. In refrigerator magnets there's certainly energy there and it can exert a force over a distance by picking up objects. You also have potential energy in a stretched rubber band. You can take a rubber band and pull it back, and there certainly is potential energy here, and you could let it go, and off it flies. So, there are all kinds of potential energy that we use in our everyday life.

The third kind of energy is wave energy. Waves represent a very efficient kind of kinetic energy in which energy can be moved long distances without actually having to move lots of mass from one place to another. There are two kinds of waves that accomplish this feat in slightly different ways. You have transverse waves, and these are sort of like the waves in a sporting event at a stadium where people stand up and then they sit down, stand up and sit down. The people are moving up and down, the wave appears to move around the stadium, and so the direction of the wave is at right angles to the direction of the movement of the individual people, or the particles involved. That's very much like the waves on the surface of the ocean, when you see ocean waves moving along. You also have compressional waves. Compressional waves would occur, if you could imagine a whole bunch of people being crowded together and one person sort of stumbles and leans into, and that wave gets pushed through from person, to person, to person. So, that's where the direction of the wave is the same as the direction of the motion of the individual pieces of the object.

You can exhibit some of these properties of waves just with a telephone cord, and if you have one of these long cords, and you are sort of on a boring phone conversation, you could start playing with it, because you can learn some of the characteristics of waves. For example, waves have velocity. Now the chord itself is not moving from one place to another, but you can see the motion of the wave going from one end to another. You could imagine, if I had a domino, or something balanced on the end of this telephone wire, and did that, it would flick off. I could exert a force over distance, even though I'm not moving mass from here to there. I'm not doing anything other than just flipping and sending a wave going through.

Waves can travel through water, as in the ocean. They can travel through sound, that's through the atmosphere. They can also travel through solid materials. Earthquakes are waves that can transfer tremendous amounts of energy through the solid Earth. So there are lots of different kinds of waves, and we'll meet several of those throughout this course.

Kinetic energy, potential energy, and wave energy—these are all pretty obvious ways of exerting a force over a distance. But once scientists understood the definition of energy, they began to see some other subtle kinds of energy. And I want to tell you about these three other kinds of energy that finally were identified. These are heat, light, and mass.

The nature of heat was a matter of intense debate for centuries. On the one hand, there were supporters of the caloric theory; this was often associated with the influential French chemist Antoine Laurent Lavoisier, who lived from 1743 to 1794. Let me tell you about Lavoisier, a fascinating character in the history of science. He was born in 1743, the son of a prosperous French lawyer. He lived in Paris and received an excellent education. He rose to a position of power and prestige in pre-Revolutionary France. He turned from his expected career in law to science, and he became enthralled with chemistry and with mineralogy. And he used part of his family fortune to construct and equip wonderful research laboratories in Paris, France, and in a nearby suburb villa that he had. Lavoisier was the champion of liberal social reform, and he worked within the existing political system to try to effect those reforms. During the French Revolution, however, he was targeted as a sympathizer of the former regime and he was executed by guillotine on May 8, 1794. The following day, the great French mathematician Joseph-Louise Lagrange said of Lavoisier: "It took only an instant to cut off that head, and 100 years may not produce another like it."

Lavoisier's scientific work was wide-ranging, but he was best remembered for all sorts of meticulous chemical studies which involved careful documentation of both the products and the reactants of various chemical processes. This was unusual in his time to look at the products and the reactants both and weigh them out and carefully analyze all of them. He was especially interested in the chemistry of burning. He played a major role in the discovery of oxygen, as well as the role of oxygen in combustion and in oxidation reactions such as rusting. In this work Lavoisier espoused this caloric theory, which describes heat as a massless fluid, a fluid that could flow from one object to another. And if you think about the way candles behave or fires behave, this isn't such an odd idea. Candles and other combustibles appear to consume their store of caloric and just evaporate away. If you heat certain objects they seem to expand, sort of like a waterlogged log expands when it soaks up water. Perhaps objects expanded when they soaked up caloric, this fluid.

Other scholars disagreed. They saw heat as a manifestation of motion at the atomic scale, and thus they thought of heat as a mechanical property of matter. This debate was ultimately resolved by a very unlikely figure. The opportunistic American-born inventor Benjamin

Thompson, who lived from 1752 to 1814, who called himself and granted himself the title of Count Rumford.

Well, Rumford led a remarkable adventuresome life. At the age of 19 he married an extremely wealthy widow 14 years his senior. The person was said to be the wealthiest widow in all of Massachusetts at the time. He settled in Rumford, Massachusetts, and gave himself this title of Count Rumford. He became a Tory spy. That was during the American Revolution, but he had to flee to England. He abandoned his wife and his infant daughter in 1775, that's when his position in the colonies became untenable and he was about to be arrested for his work as a spy. Later in his turbulent life, he was forced to flee England, one again on suspicion of spying, this time for the French. He also had the distinction of marrying Lavoisier's widow. Remember Lavoisier, well, he married that widow, who was also very wealthy. That marriage failed miserably, with much recrimination on both sides—a turbulent life for Count Rumford.

Rumford was always fascinated by the phenomenon of heat, and he conducted numerous experiments to try to improve the use of heat in everyday life. Indeed, his invention of the modern style cook stove, and the modern fireplace gained him fame and for a time a great degree of wealth for his own inventions. At one point in his life he was appointed commandant of police in Bavaria of all places, and one of his duties there was to supervise the boring, the construction of cannon. The way you make cannon is you poured a mold of brass, and then you actually had to drill out the bore of the cannon. This was a cacophonous process, and it resulted in lots of heat and you can imagine the process of taking a huge drill and turning the barrel of the cannon and keeping the drill bit fixed. You had solid bronze barrels, and as you drilled through, you generate lots of heat. What Rumford noted was, I suppose, the unremarkable fact that when you had a dull drill bit, you produced a lot more heat than if you had a sharp drill bit. But if you think about the caloric theory, if you had a sharp drill bit, you're actually cutting out more of the bronze, and so you'd think you might produce more heat with the sharp drill bit, but no, in fact, if you had an extremely dull drill bit, and you didn't drill anything at all, you just kept grinding and grinding and grinding, you'd produce the maximum amount of heat.

So, Rumford realized that the heat was produced by friction, which is mechanical action. And so he proceeded to study this phenomenon, he said: "I was led into these investigations by accident, and in some measure against my will." I'm not sure what he meant by that. In any case, in 1797 he designed a clever experiment in which he carefully insulated the barrel of a cannon, and purposely drilled it with a very dull drill bit in a tank of water. Here's what he reported. This is a quote from one of his scientific publications. "At two hours and thirty minutes the water actually boiled. It would be difficult to describe the surprise and astonishment expressed on the countenances of the bystanders on seeing so large a quantity of cold water heated, and actually made to boil without any fire." This experiment clearly showed that heat was a consequence of mechanical work. I think you can convince yourself of this if you just rub your hands together. If you do this for a while, your hands start getting very, very hot, and you really can't keep it up for too long because of the heat generated just by fiction.

The caloric theory was buried forever when English chemist Humphry Davy did a very similar kind of experiment. He was a brilliant public lecturer. He succeeded in melting ice on a cold winter day in London, just by rubbing two pieces of ice together outside. I think it's worth taking a minute to tell you about Humphry Davy, who was one of the greatest lecturers in science of all time. A real person to be revered for someone in my profession.

He was born in 1778 in Penzance, on the coast of Cornwall. He was born in a middle class family, but he had a solid education. He was drawn apparently to poetry and literature, though he also loved nature and mineral collecting. Mineral collecting was a great thing to do in Cornwall because of all the old mines, the tin mines and so forth in that part of the world. He must have developed quite a mineral collection.

The untimely death of his father in 1794 forced Humphry to take up the apothecary trade, which led to more formal exposure to chemistry. He read assiduously. He set up his own modest laboratory to perform chemical research, and he gained quite widespread attention after the publication in 1800 of his book *Researches: Chemical and Philosophical*. Based on this study, Benjamin Thompson, that is Count Rumford, invited Davy to become a lecturer on chemistry at the Royal Institution of Great Britain, and it was there that Davy established his reputation as one of the most popular lecturers in the history of science. We're going to learn more about Davy's work in chemistry in Lecture Twenty, but in any case, what Davy did was develop a flair for kinds of public demonstrations. So when there was this cold icy day in London, he used the occasion to show that you can melt ice simply by friction. He turned it to his advantage. He was always thinking about the world in that way.

Yet another convincing demonstration that heat is a form of energy is provided by heating water in a closed cylinder with a piston. Now think about this experiment. You have a cylinder that's closed. You have water in that cylinder, and you have a heavy weight sitting on top of it. If you start heating up the water and it boils and it expands, just the expansion of the water and the formation of steam is going to raise the weight against the force of gravity. You're exerting a force over distance. So just by applying heat at the bottom end of the system, you do work. So that's another demonstration that heat has to be a form of energy.

By the way, we now know that heat is actually a form of kinetic energy at the atomic level. Heat represents the motion of individual atoms, and we'll learn a lot more about atoms a few lectures from now, but just the idea that this is really a form of kinetic energy at the microscopic or submicroscopic scale is fascinating.

There are other forms of energy besides kinetic energy, potential energy, wave energy, and heat. We also have radiation—light that travels 186,000 miles per second, that's about 300,000 kilometers per second. Think about the Sun, the energy that you get from the Sun. You go out on a sunny day and you can feel the warmth of the Sun as it comes down. That energy is traveling through the vacuum of space, so it has to be a form of energy different from the ones we've talked about so far. Indeed, the Sun is the principal source of energy at the Earth's surface. It's a form of light energy. When you see light, when you turn on a flashlight, for example, the light is coming to you at 186,000 miles per second. It's a form of energy that travels through all sorts of medium, including a vacuum itself.

And finally, we have one of the defining discoveries of the modern age of science. Albert Einstein's recognition in 1905 that mass itself is a form of energy. This is according to the very familiar equation $E=mc^2$, where c is the speed of light—300,000 kilometers per second. This means that the quantity of energy in everyday objects is unimaginable. You can do a calculation. For example, a half kilogram, a one pound loaf of bread, for example, how much energy is associated with the mass? Just plug into the equation $E=mc^2$. You have the mass—half a kilogram, which isn't very large—but the speed of light, 3 times 10^8 m/s, and then you have to square that quantity, which gives the huge number of 4.5 times 10^{16} joules in just one loaf of bread. While most of this energy is forever locked up in the mass around us, scientists have discovered special circumstances where they can convert mass to energy. And we're going to talk about that a lot more in the chapters on nuclear energy in Lecture Twenty-Two, for example.

Let's summarize these ideas about energy. Energy is a subtle concept. Its characteristics are neither intuitive nor obvious. To define energy, you first have to define work, which is a force acting over a distance. Energy then, is defined as the ability to do work, the ability to exert a force over a distance. Once you define energy, you can start cataloging the different kinds of energy about us. There are three obvious kinds of energy: kinetic energy, that's the energy of motion; potential energy, which is stored energy in various forms (gravitational or elastic or chemical), that's energy just waiting to be used; we have wave energy that transfers through various mediums (solids, liquids, or gases); and then there are less obvious forms of energy, things like heat and light and, ultimately, mass as a form of energy.

Now that we've defined energy and catalogued these various kinds, the stage is set to present the two great laws related to energy—the laws of thermodynamics.

Lecture Eight
The First Law of Thermodynamics

Principle: *"Energy may change forms many times, but the total amount of energy in a closed system is constant."*

Scope:

As scientists attempted to catalogue the different forms of energy, they also explored its behavior, such as its ability to change from one form to another. Energy constantly changes forms all around us, all the time. Intense study of such transformations has led to countless useful devices, including water wheels, steam engines, light bulbs, and the automobile. In each of these technologies, a "chain of energy" leads back to the Sun—the ultimate source of most energy at the Earth's surface.

The first law of thermodynamics states that the total amount of energy obtained by adding up all the different sources is a constant. The first convincing demonstration of the first law was devised by James Prescott Joule (1818–1889), who heated water by agitating it with a weight-driven paddle wheel. To many scientists, the first law carries a profound significance about the underlying symmetry of the natural order.

Outline

I. Scientists hold a firm belief that the universe possesses deep order. An extension of this conviction holds that matter is neither created nor destroyed. Once the many forms of energy were catalogued, scientists wondered whether energy, too, might be conserved.

II. As an illustration of nature's myriad energy transfers, we can follow the chain of energy of everyday events.
   A. Riding a roller coaster provides a classic example. An electric motor converts electrical potential energy into gravitational potential energy as the coaster slowly climbs the first hill. As the coaster crests that hill and starts its downward plunge, gravitational potential is converted to kinetic energy, and so on.

- **B.** Your car stores chemical energy in the form of gasoline, which represents ancient energy from the Sun. That energy eventually becomes kinetic energy of the car and heat in the engine.
- **C.** One of the greatest challenges in modern technology is to find cheap and efficient ways to convert one form of energy into another.

III. The first law of thermodynamics states that the total amount of energy in a closed system is constant. Nevertheless, it was difficult to demonstrate because it is difficult to devise a closed system.
- **A.** The first universally convincing demonstration of the first law was an elegant experiment devised by the English brewer and sometime physicist James Prescott Joule, who is honored today by the unit of energy, the joule.
  1. Joule agitated water with a paddle wheel, which was turned by a clever linkage to a descending pair of heavy weights. He then compared the gravitational energy released by the weights' descent to the increase in water temperature.
  2. This phenomenon also occurs in a waterfall; the higher the waterfall, the greater the heating.
  3. The first law of thermodynamics allows the precise mathematical description of energy transfer events.
- **B.** The first law of thermodynamics says nothing about how fast energy is transferred. A candle and a firecracker may store the same amount of chemical potential energy, but the two objects behave very differently when lit.
  1. The rate of energy release, called power, is defined as energy divided by time. The standard unit of power is the watt, which equals one joule per second. The watt is named after Scottish inventor James Watt (1736–1819) who contributed to improvements in the Newcomen steam engine and who coined the term "horsepower."
  2. The concept of power comes forcefully into play in sports. Winners are often determined by how fast energy is released.

C. Energy is a pervasive theme in science, and every subdiscipline must incorporate the concept. Consequently, a bewildering number of energy units have been devised.
   1. Home energy bills in the United States most commonly use kilowatt-hours for electricity, therms for natural gas, and gallons for heating oil.
   2. The English system employs the foot-pound or the horsepower-hour (about 2 million times larger).
   3. Physicists use joules or ergs (one millionth of a joule) for everyday objects and events, but resort to electron volts when dealing with atomic-scale processes.
   4. Chemists often use calories in everyday chemical reactions, but they resort to a number of other units, such as the reciprocal centimeter, in special cases.

IV. To many scientists, the first law carries a profound significance about the underlying symmetry of the natural order. To Joule, the first law was nothing less than proof of the beneficence of the Creator. Others saw in the conservation of energy a natural law analogous to the immortality of the soul.
   A. British physicist William Thompson (1824–1907), known as Lord Kelvin, was drawn to the perfection of the conservation of energy, and was equally repelled by Darwin's theory of evolution, in which random variations and chance play a central role. Kelvin used the first law as a refutation of Darwin's ideas.
   1. Every closed system, such as the Earth and the Sun, has a fixed budget of energy. For life to exist on Earth, the Sun must expend that energy at a prodigious rate. Making a few simple assumptions about all known sources of energy available to the Sun, Kelvin estimated that life on Earth must be significantly less than 100 million years old—much less than the hundreds of millions of years required for Darwinian evolution.
   2. Kelvin's reputation as a physicist was so great, and the laws of physics were so unshakable, that his pronouncement was for almost half a century the major hurdle for acceptance of Darwin's theory.
   3. The conflict was resolved in 1904 when Ernest Rutherford announced the discovery of a powerful new

energy source, radioactivity—a manifestation of nuclear energy.

B. The first law provided a framework for investigating energy, but, as we shall see in the next lecture, it's only half the story.

Essential Reading:

Trefil and Hazen, *The Sciences: An Integrated Approach*, Chapter 3.

Supplemental Reading:

Burchfield, *Lord Kelvin and the Age of the Earth*.

von Baeyer, *Maxwell's Demon*, Chapters 1–3.

Questions to Consider:

1. Trace the chain of energy that allowed you to exert a force, to push a button, to start this lecture.
2. Would your life be noticeably different if a small fraction of energy vanished after every energy transfer, instead of being conserved? Why do you think scientists feel so strongly about the "truth" of the first law of thermodynamics?

Lecture Eight—Transcript
The First Law of Thermodynamics

Energy is at once an extremely useful and an extremely subtle concept. It took scientists centuries of studying heat and light and various mechanical processes before the basic definition of energy—the ability to do work—emerged. Once that concept was formulated, however, the systematic behavior of energy, the science of thermodynamics advanced quite rapidly. The first great law of thermodynamics states: Energy may change forms many times, but the total amount of energy in a closed system is constant.

I have three objectives in this lecture. First, I want to explore the many ways the different kinds of energy transform one form into another. Then we're going to look at the first law of thermodynamics—the principle that energy can change form many, many times, but that the total amount of energy is constant. Finally, I'm going to tell you some of the consequences of the first law, including the concept of power, and a fascinating story about how the first law influenced ideas and debates regarding the origin and the evolution of the Earth.

Let's begin by reviewing what we've learned already about energy. To understand energy, you first have to understand the formal concept of work, which is a force acting over a distance. Anytime an object accelerates, Newton says, there's a force involved, and a force has to be acting over a distance. So, according to Newton's first law of motion, energy is what allows acceleration, allows forces to operate in the first place. Energy is a very fundamental concept in this regard. Here we come to energy. Energy is the characteristic of a physical system that gives it the ability to exert that force over a distance. Armed with this definition, we can begin to catalogue all the different kinds of energy in the natural world. All the different ways to exert a force over a distance, and just ask yourself how you might exert such a force.

In the last lecture, we enumerated six of these different ways. You could throw something—that's kinetic energy of a moving object. You could store up potential energy—water behind a dam, for example, or a coiled spring, a battery, a magnet, something of that sort. You could generate a wave, a wave of sound, a wave in water, an earthquake. Then we also saw that heat is a form of energy—the form of energy associated with the kinetic energy of individual atoms. Light is a form of energy. It reaches us from the Sun and travels 186,000 miles per second. And finally, under special circumstances, mass itself is a form of energy that has the ability to do work. That's quite a respectable inventory. So we're now ready to look at the systematic behavior of energy, the way it behaves in the universe, and look for the universal laws that govern energy.

Let's begin though by thinking about symmetry. Symmetry is an amazingly important concept in science. Symmetry is the belief held by many, many scientists that there's some kind of balance, some kind of order in nature. Scientists of the past four centuries have held a firm belief that this universe possesses deep order. An extension of this conviction holds that matter is neither created, nor destroyed. Under ordinary circumstance, atoms don't just disappear. They don't pop into existence.

When a quantity like mass doesn't change, physicists refer to that situation as a conservation law. We say that in the closed system, matter is conserved. I'm introducing a new piece of jargon here, this idea of a closed system, and I want to make sure you get the understanding of a closed system. A closed system is like a box. And this whole concept of system is very important throughout science; it's like a box that contains matter and it contains energy. And if it's a closed system, that box is closed. No matter flows in or out of the box. No energy flows in or out of the box. And we talk about systems like this in science all the time. We have the solar system. We have ecosystems. We have nervous systems, various systems. Sometimes they're partially open. Sometimes they're closed, depending on how you want to define it. But, this first law of thermodynamics relates to closed systems.

An open system is one in which matter and energy can flow in and out. That would be true of most of the systems in your body, and also true of ecosystems, of course; that's a system in which sunlight can come in and matter does flow in and out. So we will be referring to systems quite often.

Matter is conserved in a closed system. Matter is neither created nor destroyed, and there are other various quantities that are conserved also in nature. We saw in Lecture Five that linear and angular momentum are conserved. So, what about energy? We catalogued many different kinds of energy, and scientists ask the question, are these different kinds of energy also conserved? So, how would you test something like that? How would you go about testing the conservation of energy? Well, you'd have to develop first a deep understanding of the different kinds of energy, and the way that energy can change from one form into another. Since we have so many different forms, you'd have to weigh up all these different forms in your inventory of energy to see if energy is conserved, and that's exactly what scientists of the 19th century did.

As a myriad of nature's many different ways of transforming energy, we can look at a few everyday events—sequence of events that we see all the time. One of the more dramatic ones is riding a roller coaster. This is a great way to think about transferring energy from one form to another. You start in the roller coaster, you're sitting still, then all of a sudden an electric motor starts cranking and bringing the roller coaster higher, and higher, and higher up the hill, and you feel that rattling as you're going up the roller coaster, and you get more and more gravitational potential energy in a situation like that. You crest that very first hill, and then as you start going right over the top of the hill you start accelerating. You get more and more kinetic energy as you plunge down, and of course, you have less gravitational potential energy in the process.

Some of the kinetic energy is converted into sound—the shrieking sound of the wheels. Some of it is converted into the heat of friction as the wheels hit the rails, as the axles grind against the chambers in the cars themselves. As the roller coaster goes up and down, energy constantly shifts back and forth between kinetic and gravitational types of energy. Ultimately, though, all the roller coaster's energy is dissipated as heat—heat in the rails, heat in the brakes of the roller coaster, and so forth.

You see other ways that energy transfers back and forth. You can do the same sort of thing if you drop an object. You can drop a ball. It has gravitational potential energy; it ends up on the table or the floor. Sometimes the conversion of one form of energy to another is more efficient. A tennis ball bounces a little bit more because it has more elastic energy, and at various stages while it's bouncing, it has more elastic potential energy. If you have a super ball, which has an extremely high elastic modulus, it bounces much, much higher because there's a better, more efficient conversion of the kinetic energy into elastic potential energy, so it continues to bounce more and more. So we see different kinds of conversion—potential energy of the gravitational sort, of the elastic sort; kinetic energy; and so forth, back and forth conversioning over and over again. Another example is provided by the famous Newton's cradle, in which you see constant switching of motion of one sort into another. The energy seems to be constantly being transferred one place to another, gradually as these balls hit each other, they're also heating up slightly. You also have friction in the various parts of the machine, but that's another example of the transfer of energy from one form to another.

Your car is a great example of the transfer of energy from one form to another. The first thing you do is go to a gas station and fill up with gasoline. That's a form of chemical potential energy. That energy is converted to heat in your engine, that helps drive the engine. The heat is converted into kinetic energy. The kinetic energy then drives your car. It also can be converted to electrical energy in the generator in your car, which can power your radio. That can be turned into light energy, for example, when you turn on the headlights of your car, or sound energy if you turn on your radio. There are all different kinds of energy transfer going on constantly. Ultimately, you put your foot on the brake, the brakes get hot, and some of the energy that you've paid at the gas pump goes into heating up your brakes.

You might think also backwards from the gasoline. Gasoline is refined from crude oil. Crude oil is a kind of fossil fuel; it's pumped out of the ground, and it represents decayed plant matter that was deposited in Earth millions of years ago. That plant matter got its energy from the Sun's radiant energy. So, in a sense, you can convert all the way back from the motion of your car, all the way back to some time millions of years ago, when the Sun's energy shone down on the surface of the Earth and plants created plant matter, chemical potential energy from the Sun's energy. The cycle goes on and on.

One of the greatest challenges of modern technology is to find cheap and efficient ways of converting one kind of energy to another. Power plants convert chemical energy of coal, gravitational potential energy of water, sometimes the kinetic energy of wind, even nuclear energy of mass into producing electricity. And you want to do that as efficiently as possible, because energy is expensive. Modern appliances are designed to produce kinetic energy in sound, light, and heat from electricity. So that's another thing, we are constantly are looking for ways of converting one form of energy into another.

You can play this game. You can name almost any two forms of energy. You can list all those different forms of energy, all the different kinds of potential energy, gravity, chemical and magnetic, and so forth. You can list them in one column, and also in a horizontal row, and you can match them up. How do you convert one kind of energy to another? You could play this game with your kids. You can actually right down on index cards all the different kinds of energy, shuffle the deck, and have your child grab two of them and then try to think how would you convert one kind of energy into the other. And almost always there's a technology that will do that. Sometimes they're pretty subtle, but there are technologies that we have now to convert one kind of energy into another.

Okay. So energy can be converted from one form into another. There are lots of ways to do it, but the question then is is energy conserved? That's a very different question. Is the total amount of heat and potential energy and kinetic energy, is that constant through all these different transformations? And you have to then start quantifying energy. You can't just talk qualitatively about the amount of energy in a roller coaster; you have to quantify it. How much heat energy is contained in a pot of boiling water? How much gravitational potential energy is there in suspended weight? I mean how much energy in joules? How can you calculate that? These calculations require a much more quantitative way of thinking and understanding energy.

The first law of thermodynamics states categorically that the total amount of energy in a closed system is constant. No matter how many times it changes form, energy is neither created nor destroyed. That's the statement of the first law. Once that concept of energy began to be understood and all these different forms of energy began to be studied and quantified, then scientists suspected that this law must be true. But they weren't sure; they had to perform experiments of various sorts. But it's extremely difficult to demonstrate. It's difficult because it's so hard to devise an experiment within a perfectly closed system. Energy always tends to leak in or leak out of a system.

The first universally convincing demonstration of this first law was an elegant experiment by the English physicist, James Prescott Joule.

He is honored today by the unit of energy, the joule, which we've talked about. Joule was born in 1818. He was the son of a wealthy Manchester brewer. He was home educated, and had the best education possible with home tutors. He never held a professional job throughout his entire. Joule's scientific work was supported by the family fortune, and he spent a great deal of his time and his energy trying to improve the efficiency of electric motors, which were just in their infancy when he began doing this research. In this effort, he was extremely concerned with accurate measurements of the conversion of energy from one form to another, for example, the conversion of electrical energy into mechanical work.

In his most famous experiment, Joule agitated water with a paddle wheel. The paddle wheel was turned by clever linkages to two heavy weights, which were suspended by gears and pulleys. So as the weights descended, the paddle wheel turned. And he had an insulated chamber filled with water. What he did then was measure the temperature of the water as the paddle wheel agitated the water, and if you believed in the first law of thermodynamics, as the gravitational energy of the weights descended decreased, that energy had to be transferred into the heat of the water itself. He built very beautiful and accurate thermometers. He was able to measure differences in water temperature as you agitated the water with the paddle wheel, and he measured the direct conversion of 772 foot-pounds of mechanical energy, that is, the suspended weights into one British thermal unit of heat energy—that is, the energy required to raise the temperature of a pound of water by one degree. These are the archaic units that Joule used, so he actually saw the mechanical equivalent of heat in these experiments.

This phenomenon also occurs in a waterfall. The higher the waterfall, the greater the heating of the water. As it descends gravitationally—the water plunges down and crashes into the rocks below—the water actually heats up. There's a story, although it's probably apocryphal, that in 1847 Joule was on his honeymoon, and he and his young bride were hiking in the Swiss Alps, and they actually carried out this measurement. He carried thermometers with him, he stuck them in the stream above the waterfall and below the waterfall, and he supposedly measured the difference in temperature of the river at those two spots—probably not true, but it makes a great story.

It's the first law of thermodynamics that allows you to precisely calculate the mathematical change of energy from one form to another. I want you to consider just one example to give you a sense of the mathematical power of this law. Consider the conversion of gravitational potential energy to kinetic energy of a 40-meter-tall roller coaster. So you have a roller coaster that's 40 meters tall, and how much kinetic energy can you actually get out of a roller coaster like this? Well, gravitational potential energy of an object of mass, m, raised to a height, h, recall, is mass times the gravitational constant times h—m×g×h. So when you dropped that potential energy, it is completely converted to kinetic energy, which is $\frac{1}{2}mv^2$. So we have the equation $mgh = \frac{1}{2}mv^2$. Gravitational potential energy gets converted into kinetic energy.

So, what's the velocity, the maximum possible velocity of the roller coaster that's 40 meters tall? You can calculate that out. First thing you can do is you can multiply each side by 2 over m, and simplify the equation. That tells you that v^2, the velocity squared, is equal to 2 times the acceleration due to gravity times the height. So substituting the number, height is 40, 40 meters, gravitational constant is 9.8 m/s^2. So you can substitute that v^2 is 784 m^2/s^2. Taking the square root of both sides, the velocity, the maximum velocity, is 28 m/s. That means the roller coaster's top speed is about 60 miles per hour.

You can see this kind of energy transferred all sorts of ways. Anytime you drop a weight, what you're doing is you're converting gravitational potential energy to the heat energy of the floor. The floor is now a little bit warmer than it was just a couple of seconds ago. It's hard to measure that change in temperature, but it's true.

But there are other experiments of this sort you can do yourself with your kids. One I particularly like is if your kids are feeling very rambunctious, very energetic, and maybe it's a rainy day and you need to give them something to do that's energetic, give them a jar filled with sand. Let them measure the temperature of the sand with a thermometer. Seal up the jar and then tell them they have to shake the jar until they raise the temperature by one degree. They're converting mechanical energy into heat. It's going to take them awhile to do this and they're going to use up some energy, but they'll keep opening the jar and measuring the temperature and they'll have to seal it up and keep going. It's a great activity and it's a way of demonstrating the first law of thermodynamics, just with a jar of sand.

At this point I want to introduce another common term, and that term is power. It has a very specific meaning in science. The first law of thermodynamics says nothing about how fast energy is transferred from one form to another, but your common sense tells you that this rate of energy release is very important. An example, a candle and a firecracker may transfer the same amount of chemical stored energy into heat and light, but they do it at very, very different rates when they're lit. The candle burns gradually, and the firecracker explodes all at once. This rate of energy release is called power, and it's defined as energy divided by time. The standard unit of power is the watt, which equals one joule per second. This unit of power is named for the Scottish inventor James Watt. He lived from 1736 to 1819, and he made fundamental improvements in the steam engine. I want to tell you about James Watt. He's a great character in the history of technology. He was born in Greenock, Scotland, in 1736. He was always mechanically adept, apparently, and at the age of 18 he went to London to study instrument making.

Upon his return to Scotland, he was appointed mathematical instrument maker for the University of Glasgow. So here he was a technical person. He was a technician. He wasn't a scientist, per se, but he made a fundamental important discovery. Watt began thinking about the design of steam engines when he was given the university's model of a primitive Newcomen-type steam engine to repair. So he's repairing this model and he looked at this and said, "Boy, this isn't very efficient." In the Newcomen engine there's a steam that drove a piston, but then a spray of cold water cooled the piston back down again, so we're using cold water, then you had to heat up the water again and then you have cold water again. This process resulted in a terrible waste of energy. It was a very inefficient kind of steam engine.

Watt's great improvement was to devise a separate condenser that kept the water warm, that you were boiling, and reduced fuel consumption drastically. Watt, throughout his life, made many other improvements in the steam engine, and he continued studies in the steam engine until his death in 1819. Watt is remembered today, primarily, for his unit of power. We also commonly talk in electrical power supply has things like kilowatts, that's thousands of watts, and so forth. So, you hear about kilowatts and megawatts. So we honor Watt with the unit of power.

But, ironically, Watt himself introduced a very different unit for power, that was the horsepower. The horsepower is 540 foot-pounds per second. The reason he did this was that he was trying to sell steam engines and in the early days it was very difficult to convince mine owners to buy a steam engine to replace the horses. So he said, "my engine is the equivalent of so many horses," and that's why he used this idea of the horsepower.

The concept of power comes forcefully into play in sports. It's not just how strong you are, but it's how fast you can release your energy that determines who the winner is going to be in many sports. It's true for pitchers in baseball; it's true for homerun hitters, power hitters. It's true for sprinters and shot putters. It's certainly true for boxers and running backs. It's that burst, that fast release of energy, that gives you the most efficient use in many sports.

We've defined power as energy divided by time, that is, P is equal to E divided by t ($P = E/t$). This gives us a different way to calculate energy because you can move this equation and say Energy is equal to power times time ($E = Pt$). So, for example, this allows us to calculate the energy used by a 100-watt light bulb in 10 hours. Plugging into the equation $E = Pt$, energy is the 100 watts times 10 hours, or 1 kilowatt-hour. That's what you pay for in your electric bill, and if you want to cut down on your electric bill, you have two choices—you can either turn off the lights, thus lowering the time that the light bulb is on, or you can use a light bulb of lower wattage, thus cutting down the amount of power that is consumed by that bulb.

Energy is a pervasive theme in science, and every subdiscipline has to incorporate the concept of energy. Consequently, there have been a bewildering variety of energy units that have been introduced into science. You probably have used half a dozen different energy units just in your day-to-day experience. And all these units are interchangeable. They all stand for energy. They all can be converted to joules with appropriate scaling factors. For example, home energy bills in the United States: they commonly use kilowatt-hours for electricity, they use therms for natural gas, gallons for heating oil, you use tons of coal, you use cords for wood. In other countries they sell energy by the British thermal unit, the BTU. The English system employs the foot-pound, or the horsepower-hour, that's about two million times larger. And you might think that scientists would get away from these confusing units, that this would be something that really scientists would want nothing about, but it's just not true. Physicists, they use joules, but they also use ergs, which is one millionth of a joule, for everyday objects. They all use electron volts as an energy source, or an energy term when dealing with atomic scale processes. These are all just different units for energy.

Chemists, then, use another whole set of units, reciprocal centimeters, in some cases. Calories, and actually the calorie that chemists use is one thousandth of the calorie that you see in food substances, which is the energy contained in food, so the calorie in your food is actually the kilocalorie of the chemist. This gets really confusing at times. And there're many, many other odd energy units as well. The situation is just as bad, by the way, for units of force, for mass, for power, and so forth. Units are a big problem in science and to systematize and unify them is something that we would all like to see happen. If you ever have to do a calculation in a science class, by the way, one of the key things to do is to make sure you get your units straight and know your conversion factors, and you can often find such conversion tables in the back of science text books—a really useful thing to have if you have to make those calculations.

The first law of thermodynamics is more than just a useful statement about energy. To many scientists it carries a very deep and profound significance about the underlying symmetry and the beauty of the natural order. Joule described the first law in the following poetic way. Let me read you a quote by Joule. "Nothing is destroyed. Nothing is ever lost, but the entire machinery, complicated as it is, works smoothly and harmoniously. Everything may appear complicated in the apparent confusion and intricacy of an almost endless variety of causes, effects, conversions, and arrangements, yet it's the most perfect regularity preserved. The whole being governed by the sovereign will of God." To Joule the first law was nothing less than proof of the beneficence of the Creator. Others saw in the conservation energy, in that natural law, something analogous to the immortality of the soul, and they made those connections.

I want to end by telling you about one other British scientist, the great physicist William Thompson, was also known as Lord Kelvin, and he certainly agreed with these sentiments about the beauty of the first law of thermodynamics. Kelvin was a towering figure in the Victorian era of science. He was born in Belfast, Ireland, in 1824. At the time, his father was professor of mathematics at the Royal Academic Institution of Belfast, and so young William was home taught and got the most rigorous education until the age of 10—that's when his father was offered the Chair of Mathematics at the University of Glasgow, and William, at the age of 10, entered the university. He was the top in his class in mathematics, in logic, and in classics. He published the first of some 660 scientific papers at the young age of 16. He went on to Cambridge for what was, in effect, graduate work in theoretical physics. While there he was also on the rowing team, he helped found the Cambridge University Music Society, and that's still thriving today. After a brief time in Paris, he became professor of natural philosophy at Glasgow, and that was the position he held then for 53 years until his death in 1907. He was buried in Westminster Abby near Newton and Darwin, and that should give you some sense of the esteem in which he was held.

Kelvin was drawn to the perfection of the first law of thermodynamics, the conservation of energy. He appreciated the economy, the symmetry of this law. On the other hand, Kelvin wrote with great distaste about Darwin's theory of evolution, in which random variations and chance seemed to play a central role. How could the creator of such magnificent order of the first law of thermodynamics then resort to chance and probabilities when it came to his highest creation, life? So Kelvin set out to use the first law of thermodynamics to destroy Darwin's idea. This is how he did it. He said that every closed system such as the Earth and the Sun has a fixed budget of energy. For life to exist on the Earth, the Sun has to expend prodigious amounts of energy. And this is what Kelvin said, "Within a finite period of time past, the Earth must have been, and within a finite period of time to come, the Earth must again be, unfit for the habitation of man as at present constituted." In other words, you only had a fairly short span of time for humans to evolve and to live on the Earth. Making a few simple assumptions about all the different known energy sources in his time—the Sun, the various kinds of fuels, gravitational collapse of the Sun, and so forth—he decided that life on Earth must have been around for significantly less than 100 million years, perhaps only 10 million years. And this was much, much less than the hundreds of millions, or perhaps billions of years required by Darwin and his theory of evolution.

Kelvin was very wise to add, when he said this, by the way, "unless sources of energy now unknown to us are prepared in the great storehouse of creation," so he kind of hedged his bets.

It was Kelvin's reputation as a physicist that was so great, and the laws of physics seemed to be so unshakeable, that his pronouncements were for a half a century the principal hurdle, the major impediment to the acceptance of Darwin's theory of evolution. Darwin even referred to Kelvin in his opposition as "an odious specter," because he realized you couldn't fight the first law. This conflict was resolved in 1904, when Ernest Rutherford announced the discovery of a powerful new energy source—radioactivity— that's the manifestation of nuclear energy, and we're going to learn a lot more about that in Lecture Twenty-Seven. Einstein's famous equation, $E=mc^2$ appeared the following year. So clearly, the Sun and the Earth had major energy sources that no one had recognized before, and that conflict between physics and biology was finally resolved.

To summarize, this lecture is focused on the ways that energy can change from one form to another. We've seen that energy can change form many, many times, and that the rate of change of energy, or power, can vary widely. Nevertheless, the total amount of energy in a closed system is constant. Energy is neither created, nor is it destroyed. The first great law of thermodynamics provides a glimpse of the order and the symmetry of the universe. The first law also provided a framework for investigating energy, but as we'll see in the next lecture, the first law is only half of the story.

Lecture Nine
The Second Law of Thermodynamics

Principle: *"Heat has a universal tendency to dissipate."*

Scope:

The first law of thermodynamics says nothing about the ways energy transitions can occur. We experience many limitations on energy transfers—restrictions which are summarized in the second law of thermodynamics. The second law of thermodynamics, in its most intuitive form, says that heat tends to diffuse evenly, spreading out from warmer to cooler objects. This statement distinguishes between two related concepts, heat and temperature. Heat is a quantity of energy, whereas temperature is a relative term—two objects are at the same temperature if no heat flows spontaneously from one to the other. Heat flow can occur by three different mechanisms: conduction (the transfer of heat from atom to atom in a solid object), convection (the transfer of heat through a moving fluid, either a liquid or a gas), and radiation (the transfer of heat by a form of light that travels 186,000 miles per second).

Another, more subtle statement of the second law is that an engine cannot convert heat energy completely into useful work. The French engineer Nicolas Léonard Sadi Carnot (1796–1832) derived a mathematical relationship for the maximum possible efficiency of any engine, which depends only on two easily measured temperatures: the hot reservoir of the fuel and the cold reservoir of the surroundings into which waste heat is dumped.

Outline

I. The first law of thermodynamics states that the total amount of energy is constant, but it says nothing about the ways that energy can shift from one form to another. In everyday life we experience many limitations on energy transfers. A hot bowl of soup becomes cooler, for example, but a cool bowl of soup never spontaneously heats up. A second law of thermodynamics is required.

II. The most intuitive statement of the second law of thermodynamics is that heat tends to diffuse evenly: heat flows from hot to cold.
  A. This statement incorporates two related, but different, concepts—heat versus temperature.
    1. Heat is a quantity of energy (joules, calories, etc.). The quantity of heat energy thus depends directly on the amount of material the system contains.
    2. Temperature is a relative term: two objects are at the same temperature if no heat energy flows spontaneously from one object to the other.
    3. Every temperature scale requires two reproducible reference points.
    4. At absolute zero, −273.15°C, the kinetic energy of atoms and molecules is zero.
  B. A consideration of heat versus temperature leads us to another important property of materials—the heat capacity. Every material has the capacity to store heat energy, but some substances do this better than others.
    1. Think about placing a pound of copper and a pound of water on identical burners. Which one heats up more quickly?
    2. Heat capacity is the amount of heat energy that a substance can hold. Water has a much higher heat capacity than copper.
    3. A useful measure of the heat capacity is called "specific heat," the energy required to raise the temperature of 1 gram by 1°C. For water, this amount of energy is defined as the calorie. By contrast, it takes only 0.1 calorie to raise a gram of copper by a degree.

III. The second law of thermodynamics depends on the motion of heat.
  A. Heat can move by three different mechanisms: conduction, convection, and radiation.
    1. Conduction is the transfer of heat from atom to atom in a solid object.
    2. Convection is the transfer of heat through a moving fluid, either a liquid or a gas.

3. Radiation is the transfer of heat by a form of light that travels 186,000 miles per second.
B. Insulation in animals, in clothing, and in houses is designed to reduce the inevitable transfer of heat.
   1. Most heat loss comes from convection. Fur and fiberglass insulation trap pockets of air so small that convection can't occur.
   2. Thermos bottles carry this strategy a step further by having a vacuum barrier, across which neither conduction nor convection can occur.

IV. The second law of thermodynamics applies to so many physical situations that there are a number of very different ways to express the same fundamental principle. A second, more subtle statement of the second law is that an engine cannot be designed that converts heat energy completely to useful work.
   A. Thermodynamics were of immense importance to designers of steam engines, so many theoretical insights came from this practical device. The French military engineer Nicolas Sadi Carnot (1796–1832) came the closest to deriving the second law from a study of work and heat.
   B. Carnot considered two sides of the relationship between work and heat.
      1. He recognized that work can be converted to heat energy with 100% efficiency. It is possible to convert the gravitational potential of an elevated object, or the chemical potential of a lump of coal, completely to heat without any loss.
      2. Converting heat to work is more restricted. Inevitably, some heat winds up heating the engine and escaping into the surroundings.
      3. You can expend work to raise water and fill a reservoir; with care, without losing a drop. But, if water is released to produce work, some of the water has to flow through the system.
   C. Carnot's great contribution was the derivation of the exact mathematical law for the maximum possible efficiency of any engine—that is the percentage of the heat energy that can actually do work.

1. The maximum efficiency of an engine depends on two temperatures: T_{hot}, the temperature of the hot reservoir (burning coal, for example), and T_{cold}, the temperature of the cold reservoir of the surroundings into which heat must flow (usually the air or cooling water).
2. An engine is a mechanical device imposed between two heat reservoirs. In this sense, an engine is any system that uses heat to do work—the Sun, the Earth, your body, or a steam engine.
3. Efficiency of an engine can be expressed by the formula

$$\text{Efficiency} = \frac{T_{hot} - T_{cold}}{T_{hot}} \times 100$$

4. In Carnot's day, before these principles were understood, typical steam engine efficiencies were 6%. Today, improved insulation and cooling have raised efficiencies of coal-burning power plants to 40%, which is close to 90% of the theoretical limit.

D. Carnot's equation illustrates why fossil fuels—coal, gas, and oil—are so valuable. These carbon-rich fuels burn with an extremely hot flame, thus elevating the temperature of the hot reservoir and increasing the maximum efficiency.

Essential Reading:

Trefil and Hazen, *The Sciences: An Integrated Approach*, Chapter 4.

Supplemental Reading:

Atkins, *The Second Law*.

von Baeyer, *Maxwell's Demon*, Chapters 5–17.

Questions to Consider:

1. Identify three examples of the second law of thermodynamics that are occurring where you are, right now. Can you think of any physical event where the second law does not come into play?
2. If engines operate more efficiently in colder surroundings, why don't we build all our power plants in the Arctic, to minimize T_{cold}? (There may be several reasons.)

Lecture Nine—Transcript
The Second Law of Thermodynamics

The first law of thermodynamics is wonderful news. You can change energy from one form to another and to another as many times as you want to, and the total amount of energy is constant. Energy is conserved. But nature hasn't given us a completely free ride in this regard. The second law of thermodynamics places severe restrictions on how energy can be used, how it can transfer from one form to another. In fact, the second law of thermodynamics tells us that heat has a universal tendency to dissipate.

In the last lecture, we looked rather uncritically at the many different ways that energy shifts from one form to another, from kinetic to potential to heat and back and forth, and so forth. We call that example the roller coaster. The roller coaster starts at the bottom. You take electrical energy, bringing that roller coaster up and up and up, and it gains gravitational potential energy. That changes into kinetic, to gravitational potential, and back and forth, and ultimately to heat energy in the brakes of the roller coaster as it comes to a stop at the end of your ride.

In this lecture, and in the next, my principle objective is to introduce one of the most profound ideas in science: this second law of thermodynamics. In coming to grips with this law, we're going to have to scrutinize nature, to see the restrictions on how we can use energy. Such limitations, as outlined in this second law of thermodynamics, have profound consequences for our everyday experience. The British scientist and science commentator C. P. Snow once wrote, "Once or twice I've asked a group of educated people how many of them could describe the second law of thermodynamics. The response was cold. It was also negative. Yet, I was asking something which is about the scientific equivalent of, 'Have you ever read a work of Shakespeare?'"

Well, why is this second law of thermodynamics so important? The first law, recall, states that the total amount of energy is constant, but it says nothing about the way that energy can shift from one form to another. In everyday life, we experience many limitations on these energy transfers. For example, a hot bowl of soup becomes cooler spontaneously, but a cold bowl of soup never becomes hot spontaneously. A glass of ice water gradually comes to room temperature, and the ice melts, but you never see a glass of water sitting on your table and then suddenly ice cubes form in it. Either way, energy is conserved. Yet one example is commonplace, and the other never happens.

Another example is gasoline. Gasoline burns in your car to produce heat and exhaust, but you never see exhaust plus heat spontaneously become gasoline. And yet once again, that wouldn't violate the first law of thermodynamics, because the total amount of energy in its different forms would remain constant. Another example: objects fall downward in a gravitational field. You let go of an object, it falls, and it releases heat when it hits the bottom. For example, an asteroid coming and hitting the Earth would release a tremendous amount of heat. But you never see the other thing; you never see an object falling up and absorbing heat in the process. So objects don't fall up.

I want you to imagine playing a favorite videotape backwards. Certain actions and events are going to look silly, they're going to look wrong, and they're going to make you laugh. Let me emphasize that all of the events, whether the tape is played forward or backwards, are perfectly consistent with the first law of thermodynamics, because energy is conserved either way. So, evidently there are restrictions on the flow of energy, and that's why the second law of thermodynamics is required.

The most intuitive statement of this second law comes to us courtesy of that British physicist William Thompson, Lord Kelvin, who we met in the last lecture. Kelvin said that heat tends to diffuse; it tends to move out evenly. Heat flows from hot to cold. This statement incorporates two related but different concepts, heat and temperature. And once again, we have to come into this situation of common, everyday words like "heat" and "temperature" having very specific scientific meanings.

Let me tell you about these two words. Heat is a quantity of energy. It's a measurable unit; you can measure the amount of heat an object holds in joules or in calories. Two kilograms of water at room temperature contain twice the heat energy of one kilogram of water. You can take a certain amount of water that's at a certain temperature, there's heat energy here, and you can transfer that heat energy into different receptacles, just like that. Now, some of the heat energy that was contained in this jar of water is contained in two glasses. The quantity of heat energy thus depends on the quantity of stuff and the temperature at which it is.

Here, we come to this word "temperature." Temperature is a relative term. Two objects are at the same temperature if no heat flows spontaneously from one of them to another. Two objects are at a different temperature if heat does spontaneously flow from the warmer to the cooler object. You measure temperature with a thermometer. And every temperature scale, every thermometer, requires two reproducible reference points, such as the freezing and the boiling points of water. They are commonly used. You also need to have some physical property that changes, for example, the volume of mercury at a particular temperature. So you build a thermometer that has some mercury at the bottom, and as the mercury gets warmer or cooler, the volume of the mercury changes and so the mercury moves up and down, and that gives you a record of the temperature. So a reproducible physical property is also essential.

Lord Kelvin measured the heat energy contained in substances at different temperatures. He realized that at a very specific temperature, at -273.15 degrees Centigrade—that's about 460 degrees Fahrenheit—you can't get any more heat energy out of any object. And he called that temperature absolute zero, the temperature at which there was no more heat energy that you could pull out of a material. At absolute zero, the kinetic energy of atoms is as close to zero as you can possibly get it. The Kelvin temperature scale, therefore, uses 273.15 and 373.15 as the reference temperatures, that's the freezing and boiling point of water. Note that both the Celsius temperature scale, which we used, and the Kelvin scale are both Centigrade scales, that is, there are 100 divisions between freezing and boiling of water.

When you start thinking about this difference between heat and temperature, it leads us to another important property, and that's heat capacity. Every material has the capacity to store heat energy, but some materials are much better at this than others. Think about placing a pound of copper and a pound of water on identical burners. I want you to think about this. Which one heats up more quickly? If you put your hand in the water and put your hand on the metal, which one would get burned first? Well, your common experience tells you that the metal gets hotter much faster, and this is because metals are unable to store very much heat energy. But the water is an extremely efficient store of energy, so water has a very high heat capacity. You have to put in a lot of energy before the temperature goes very high. Heat capacity, in fact, is the amount of energy that a substance can hold, and since water is so efficient, this has very important implications for climate and weather. We're going to see that in Lecture Forty-One when I'll talk more about the properties of water, but suffice it to say that water is a very efficient store of heat energy.

A useful measure of heat capacity is called "specific heat," yet another term in science. That's the energy that is required to raise the temperature of one gram of material by one degree Centigrade. For water, this energy is defined as one calorie, so here we have the definition of a calorie as the energy required to raise one gram of water by one degree Centigrade. By contrast, it only takes a tenth of a calorie to raise a gram of copper by one degree. So, copper has only a tenth the heat capacity of water.

Let's pause for a second and think about the second law of thermodynamics in a different regard. It's in part a statement about the motion of heat, how heat moves. And so we have to say something about the mechanisms, because the law doesn't tell us what these mechanisms of moving heat are. It turns out that there are three different ways to move heat. We have conduction, convection, and radiation, and I want to look at these three common mechanisms that we experience everyday.

Conduction is the transfer of heat from atom to atom in a solid object. If you take a spoon and you place the spoon in hot water, you notice that very quickly the handle becomes hot, too. That's the process of conduction; the heat conducts from one end of the spoon to the other. Conduction is a rather slow process. For example, heat left over from the formation of the Earth, the gravitational formation of the Earth four and a half billion years ago, is still slowly moving outward by conduction. You've experienced this fact if you've ever gone into a cave on a hot summer day or on a very cold winter day. You find that the cave maintains more or less the same temperature year-round. Even if the cave is only a few tens of feet beneath the surface, the temperature is maintained at a constant level, because heat conduction through the rock is relatively slow. And so the cave reaches this equilibrium.

Our second way of moving heat is called convection, and that's the transfer of heat through the motion of a whole body of a fluid, either a gas or a liquid. So this includes things like boiling water, ocean currents, or the wind. Those are good examples of movement of heat by convection. Your body regulates its temperature in large part by convection. For example, you breathe, you breathe in air and then breathe it out. You know how on a cold day, you see that frost on your breath? Well, you're regulating your temperature, or at least altering your temperature, through breathing. The circulation of your blood also is a kind of convection, and so is sweating. These are all ways of moving heat from one place to another by moving a fluid, a gas or a liquid. You also, in everyday life, have lots of mechanical conveniences, things like fans or hair dryers, and once again, that moves heat, transfers heat, or transfers cooler air from one place to another by just moving a body of air, by blowing it.

Now, there's a third way of moving heat. In addition to conduction and convection, we have radiation. That's the transfer of heat in the form of light energy or in heat energy, which moves 186,000 miles per second through a vacuum. You feel that radiant energy from the Sun, and that's a way for the Sun to transfer its heat energy, that's 93 million miles away, to us through the vacuum of space. You also feel this kind of radiant energy if you have a space heater, an electric heater, or a fireplace, and you feel that warmth even though you're not necessarily getting direct conduction or convection from that fire. You know, people also radiate heat, because you're generally warmer than your surroundings. That's why a crowded auditorium gets hot. In fact, I'm told, and I'm inclined to believe that it's true, that the average person just sitting in a chair radiates the same amount of heat as a hundred-watt light bulb. So, if you're in an auditorium with one or two thousand people, that's a lot of heat being radiated out, and that's why auditoriums can get quite warm if they're not well-ventilated.

One of the key technologies that we look for is ways to eliminate or reduce this transfer of heat. Humans, and nature as well, have devised many different ways of reducing the transfer of heat. Animals living in cold climates, for example, use insulation of various sources. They develop fur, or they have feathers. Warm wool clothing, for example a wool scarf, does a similar sort of thing. It has lots of little air pockets, and think about what an object like this would do, the pockets of air that wool form or the kinds of things that feather and fur would provide in terms of animals. What you're trying to do is cut down the efficiency of heat transfer, and the most efficient heat transfer is by convection. So if you can trap little pockets of air where convection, that cycling of air, can't proceed, then you're restricting the way heat is lost from your body to conduction, which is a much slower process. And that's how insulating materials work in your home, in your clothing, in animals, and so forth. Thermos bottles carry this process a step further by actually having a vacuum barrier. Across a vacuum barrier, you can't even get conduction; the only thing you can do is get radiation. So heat loss is an even slower process, although it still occurs, because heat always flows from warmer objects to colder objects, even in the vacuum of space.

The second law of thermodynamics is so pervasive, it applies to so many physical situations, that there are a number of very different ways of expressing the same fundamental principal, and I want to look at a variety of ways of talking about the second law. We've seen so far that heat flows from hot to cold, but another more subtle statement of the second law is that you can't design an engine that converts heat completely 100 percent to work. Think about that statement. Thermodynamics are of immense importance to the designing of steam engines, because you're talking about transferring heat. You're talking about using energy in different ways, so in designing steam engines, you're worried about transferring energy in an efficient way. So it's not surprising that some of these insights came from the study of the design and the development of steam engines.

I want to tell you about a French military engineer, Nicolas Léonard Sadi Carnot. He came the closest to deriving the second law of thermodynamics from a study of work and heat and the nature of steam engines. Sadi Carnot was born in Paris in 1796. He was the son of Lazare Carnot, who was a famous military tactician, one of Napoleon's great generals, and a minister of war under Napoleon. The younger Carnot was educated to be a military engineer in his father's footsteps, and he devoted much of his effort to understanding the operation of steam engines, which were just then coming into use and found applications, of course, in the military society.

His most important findings were summarized in a little book, a forgotten book, called *Reflections on the Motive Power of Fire*. It appeared in 1824, but it was completely ignored in his day. It was ignored for almost 20 years. That's long after his death; Carnot died in 1832. I think Carnot's great gift was the ability to see generalizations in very specific situations. For example, working with the mechanics of a steam engine, he could see general principles. This is something that he wrote: "In order to consider, in the most general way, the principle of the production of motion by heat, it must be considered independently of any mechanism or any particular agent. It is necessary to establish principles applicable not only to steam engines, but to all imaginable heat engines, whatever the working substance, and whatever the method by which it is operated." So, he saw steam engines as just a special case of a much more general kind of problem in thermodynamics.

Carnot considered two different sides of this relationship between work and heat. On the one hand, he recognized that you can convert work to heat with 100-percent efficiency. Work can be completely transferred to heat. For example, if you drop an object, if you take a ball and drop it, it has gravitational potential energy, it does work, it lands on the ground, and the work is converted completely to heat energy in the floor. Now, you may say, "Well, it doesn't seem like there's very much heat energy generated by dropping a ball," but imagine the same situation with an asteroid coming in from outer space and crashing into the Earth's surface. That asteroid basically is disintegrated, it's obliterated, and huge amounts of heat are released. And in principle, you can do that with 100 percent efficiency. All the energy that's represented in that system can be converted to heat. It's the same thing with a piece of coal; you can burn a piece of coal completely away and convert that energy into heat very efficiently.

But the flip side, that's converting heat into work, is more difficult. And this is true for at least three reasons. First and most simply, some of the heat that you introduce into your engine has to wind up heating the engine itself, or escaping into the environment, so you're wasting some of the heat. Some of the gasoline that you pay for ends up heating up the engine, it ends up heating up the exhaust. So, some of the heat energy is lost to the environment. Some of the food energy that you eat ends up just being radiated away, because we radiate 100 watts, just like a light bulb. Nevertheless, you can imagine lots of different ways to make more efficient insulation, of designing more efficient engines to reduce, and reduce, and reduce this heat loss. So you can imagine engineering away a lot of that loss. There's no intrinsic limit to how much of that loss you can engineer away, so you can be very, very efficient in that regard, but you're always going to lose some energy that way.

The second reason that some energy is always wasted is that an engine operates in a kind of cycle. For example, a piston pushes on a rod to drive the machine, and then you have to pull the piston back in order to operate it again. So you have this continuous operation, and some of the energy is gone to do work, but then you have to use some energy to reset the piston so you can keep doing work again. So any engine that operates in a cycle has this kind of problem. Even your heart that's pumping has that same kind of problem; you have to reset the heart before it can pump again. Once again, you can design an engine so that this energy loss is reduced. You can have extremely lightweight pistons, you can use good lubricants, and all that sort of thing, but still, you have to use some energy to reset the system.

But there's a third reason, and a more subtle one, and this is the one that Carnot really hit. You can't design away this third problem. The heat energy flows through the system from a hot reservoir, the burning fuel, to the cold reservoir, that's the surroundings, that's the outside. Here's the catch. No matter how cold that outside reservoir in which the heat flows into is, unless it is at absolute zero—and it can't be at absolute zero, you just can't mechanically design some system with absolute zero—some of the energy is wasted because some of the energy that you started with is still remaining in the low-temperature reservoir that heat flows out of. Everything, even ice water, has a lot of heat energy still in it. Whatever energy you start with, it has to flow to the outside, it has to flow through your engine system.

An analogy can be drawn between heat energy and the gravitational energy of water passing over a water wheel. I want you to think about this analogy because it makes it a little clearer; heat's a more abstract concept, but water you can hold in your hand. So, imagine designing a water wheel that's the maximum efficiency possible. The elevation of the water above sea level is analogous to the temperature of your fuel, and the level at the bottom of the water wheel is analogous to the low temperature reservoir where the water flows out, and you can imagine sea level as being zero energy or absolute zero if you will. So think about designing that water wheel.

First, you're going to have some water leaking away, the water is going to evaporate into the air, or it may soak into the material of which the water wheel is made. Now, you can design your water wheel with extreme efficiency so that you minimize the amount of evaporation and the amount of loss of water, the amount of leaking, by sealing all the cracks very carefully. But you're always going to lose a little bit of water; there's no way around that. Second, some of the water's energy has to be used to turn the wheel. And you have buckets that are pulling down, but then you have to bring those buckets back up to a position where they can go around again. So some of the water's energy has to be used just to raise up the buckets. Now again, you can design that water wheel to be made of extremely lightweight materials, you can have high-quality lubricants so that the wheel turns smoothly, but you're always going to use some of your water as the wheel goes around.

And this loss is exactly like the energy used to draw the piston of a steam engine back, and so forth. So there's an analogy there. But no matter how carefully you design that water wheel, it can't use the water all the way down to sea level. No matter how you design the water wheel, the water is going to spill out of the machine some place above sea level, and then the wheel is going to move around. So some of the potential energy of the water is always wasted, because the water wheel is always going to be at some level above sea level. No matter how you design that water wheel, its efficiency is going to be limited in that way. You perhaps could design a bigger water wheel, you could perhaps design it so that it gets as close to sea level as you possibly can, but there's always going to be some loss, and that's the intrinsic loss that Carnot recognized.

Carnot's great contribution was the derivation of the exact mathematical law that tells us the maximum possible efficiency for any engine. That is, the percentage of heat energy that can go in to doing work, as opposed to the percentage which is always going to be wasted just by these design considerations. It turns out that the maximum possible efficiency of any engine depends entirely on two easily measured numbers. First of all, we have T for the hot end, T of the burning fuel, T at the hot end of the reservoir. So that's the heat energy that's flowing into your system to do work. And at the other end of the system, you have T of the cold reservoir, usually called T_{cold}. That's the temperature of the material that flows out of the engine: the exhaust, if you will. Every power plant, every engine, has an exhaust of some sort, and it's the temperature of that exhaust. An engine is just a mechanical device that is imposed between the hot reservoir and the cold reservoir, just like a water wheel is a mechanical device imposed between high gravitational energy and lower gravitational energy. It's an exact analogy there.

Efficiency then can be defined mathematically as the difference between these two temperatures, the difference between T_{hot} and T_{cold}, divided by the temperature of the hot reservoir. To put that into a percent, you have to multiply it times a hundred. So here's the equation that you use for efficiency:

$$\text{Efficiency} = \frac{T_{hot} - T_{cold}}{T_{hot}} \times 100$$

Now, you can use this equation to calculate all sorts of efficiencies, the maximum possible efficiency for different kinds of power plants, different kinds of engines, and so forth. This becomes very important to understand the efficiency that is possible with modern engines. For example, think about an engine in which the hot reservoir is boiling water and the cold reservoir is ice water. You just plug in those temperatures, and they have to be temperatures in Kelvin, in the absolute scale of Kelvin, and here's what happens. The hot reservoir is at 373.15. The cold reservoir is at 273.15. Divide that difference by 373.15 and then multiply by 100, and you can get about 27 percent efficiency using boiling water and ice water. Now, think about it—that's not very much efficiency. You're only using 27 percent of the available heat energy to do work.

You can improve things if you have a higher temperature hot reservoir. So, for example, if you burn coal to produce steam, and that steam could be, for example, in a typical power plant, at about 550 Kelvin, let's do the calculation again. In this case, the hot reservoir is at 550. Typically, one uses room temperature, which is about 300 Kelvin, as the cold reservoir; you divide that by 550; and you see that for a coal-burning power plant, you typically can have efficiencies of 45 or 46 percent. That's the maximum possible for a coal-burning plant.

You could design things differently, for example, you could put your coal-burning power plant in the arctic, where the average temperature is only about 250 K in the outside air, so therefore your cold reservoir is colder, you're stretching out the total difference in temperature, and let's see what happens there. (550–250)/550 gives you almost 55 percent efficiency, so just by moving your power plants to the arctic, you could gain a great deal of efficiency in a coal-burning power plant. And then think about what would happen if you have an engine in deep space. In deep space, if you shield from the Sun, you might have a cold reservoir as cold as 50 Kelvin, and here you really gain a lot. Do the calculation one more time. [(550–50)/550]×100: that's over 90 percent efficiency possible in deep space, because the cold reservoir is so cold, it's getting closer to absolute zero. So in that case, you can be much more efficient. You can convert more of your heat energy into work.

In Carnot's day, before these principles were understood, typical steam engines had an efficiency of only 6 percent, believe it or not. But when people realized that they could get much higher efficiencies and they saw Carnot's reasoning, they began to design improvements, better insulation, better materials, tighter-fitting steam hoses, and so forth, and today with these improvements, efficiencies have been raised to about 40 percent, for example, in coal-burning power plants. That's close to 90 percent of the theoretical limit, so we're getting very close to the maximum theoretical efficiency of a coal-burning power plant. Here's a clear case where theoretical ideas and understanding have led to dramatic benefits for society, because we use our energy more efficiently.

Still, remember that 60 percent of the coal's energy that we burn is wasted. It's just dumped into the environment; we can't use it. Fuels can be graded according to the heat they contain. Carnot's explanation illustrates why some fuels, hot-burning fuels like coal, gas, and oil, for example, are all very much valued. It's because they burn at such high temperature. You get that bigger spread in temperature, and therefore you can have more efficient plants. About 90 percent of all the industrialized world's energy comes from fossil fuels. It's still possible to scrape coal off the ground in some places and to drill shallow wells, so for the time being, these fossil fuels are going to be very valuable and play a very important part in the world's energy budget. These fossil fuels represent ancient life forms that have been buried and transformed by the Earth's heat and temperature, and so they're limited resources. We are mining these out. Someday, we're going to use up the remaining store of fossil fuels. They are non-renewable resources. So we're going to have to find other sources of energy, and let's hope that they burn at high enough temperatures that we can keep that high efficiency that we now enjoy.

To summarize this lecture: two great laws of thermodynamics systematize the behavior of energy—that's the capacity to exert a force over a distance. The first law says that energy can change forms many, many times, and the total amount of energy is conserved. That's the good news. But the second law places limits on how this energy can transfer from one way to another. Heat tends to spread out. It flows from hot to cold. That gives a direction in the use of energy. Similarly, an engine that converts heat to work cannot operate with 100 percent efficiency, because some of the heat must escape into the environment and some has to be used to reinitialize the engine. These physical realities dictate how we can use energy. We can't escape the second law. But the second law goes much, much deeper than that. In the next lecture, we're going to explore one of the most fascinating, profound concepts in all science: the concept of entropy.

Lecture Ten
Entropy

Principle: *"Every isolated system becomes more disordered with time."*

Scope:

The discussion of the second law has focused, thus far, on the behavior of heat energy: "heat flows spontaneously from warmer to cooler objects," and "all engines that convert heat to work are less than 100% efficient." These ideas are useful, but the second law has a much wider significance; in its most general form, it states that the degree of disorder, or entropy, of any system tends to increase with time. This more general approach to the second law was proposed by Rudolf Clausius, who introduced the rather abstract concept of entropy—the ratio of heat energy over temperature. Clausius observed the behavior of steam engines and realized that this ratio must either remain constant or increase; entropy tends to increase.

A more intuitive approach treats entropy as a measure of a system's disorder—all systems left to themselves tend to increase their disorder. The second law of thermodynamics has deep and far-reaching consequences such as the direction of time—the direction of increasing disorder. Yet, in spite of the success of the first two laws of thermodynamics, profound questions remain about the remarkable tendency of the universe to spontaneously form highly ordered local systems, such as life.

Outline

I. The discussion of the second law of thermodynamics thus far has focused on the behavior of heat energy: heat flows spontaneously from warmer to cooler objects, and an engine cannot be 100% efficient. As useful as these ideas may be, the second law reaches far beyond the concept of heat. In its most general form, the second law comments on the state of order of the universe.

II. All systems in the universe have a general tendency to become more disordered with time.
  A. Many experiments can be designed to study this phenomenon.
    1. Shuffle a fully ordered deck of cards and it becomes disordered—never the other way around.
    2. Shake a jar of layered colored marbles and they become mixed up—never the other way around.
  B. These examples are analogous to a situation where a hotter and colder object come into contact. The heat energy of the hotter material gradually spreads out.
  C. In each of these cases the original state of order could be recovered, but it would take time and energy.

III. The concept of entropy was introduced in 1865 by the German physicist Rudolf Clausius (1822–1888) to quantify this tendency of natural systems to become more disordered.
  A. Clausius synthesized the ideas of Carnot, Joule, and others and published the first clear statement of the two laws of thermodynamics in 1850. The second law, however, was not presented in a rigorous mathematical form.
    1. Clausius realized that for the second law to be quantitatively useful, it demanded a new, rather abstract physical variable—entropy. He defined entropy purely in terms of heat and temperature: entropy is the ratio of heat energy over temperature.
    2. Clausius observed the behavior of steam engines and realized that this ratio must either remain constant or increase—that is, the heat divided by the temperature of the cold reservoir is always greater than or equal to the heat divided by the temperature of the hot reservoir (or heat divided by work). Thus the entropy of a system does not decrease.
  B. One can summarize the two laws of thermodynamics as energy is constant (first law), but entropy tends to increase (second law).

IV. A more intuitive approach to entropy is obtained by thinking about heat energy as the kinetic energy of vibrating atoms. Heat spreads out because faster atoms with more kinetic energy collide with slower atoms; eventually, the kinetic energy averages out.

    A. Ultimately, the order of any system can be measured by the orderly arrangement of its smallest parts—its atoms.

        1. A crystal of table salt with regularly repeating patterns of sodium and chlorine atoms is highly ordered. A lump of coal, similarly, has an ordered distribution of energy-rich carbon-carbon bonds.

        2. Dissolve the salt in water or burn the coal and energy is released, while the atoms' disorder—their entropy—increases.

    B. The "arrangement" of atoms can also refer to the distribution of velocities of particles in a gas—i.e., its temperature. Imagine what happens when two reservoirs of gas at different temperatures are mixed. Temperature averages out and the entropy—the randomness—increases.

        1. This definition of entropy was placed on a firm quantitative footing in the late 19th century by the German physicist, Ludwig Boltzmann (1844–1906).

        2. Boltzmann used probability theory to demonstrate that, for any given configuration of atoms, entropy is related to the number of possible ways you can achieve that configuration. The most probable arrangement is observed. Entropy is the microscopic manifestation of probability.

V. The second law of thermodynamics has far-reaching consequences.

    A. One consequence of the second law of thermodynamics is that heat in the universe must eventually spread out evenly.

        1. Life on Earth is possible because the Sun provides a steady source of energy, and the Earth holds a vast reservoir of interior heat.

        2. In the 1890s, many scholars addressed the concept of the "heat death" of the universe.

- **B.** The second law of thermodynamics provides us with tantalizing insights about the nature of time.
  1. Imagine playing a favorite movie backwards. Some forms of motion, such as a ball flying through the air, seem completely reversible.
  2. Other images are obviously impossible in the real world.
  3. The tendency of a system's entropy to increase defines the arrow of time.
- **C.** Living things must obey the laws of thermodynamics, which come directly into play with the concept of trophic levels.
  1. Every organism must compete for a limited supply of energy. Plants, in the first trophic level, get their energy directly from the Sun. Herbivores, the second trophic level, obtain energy from plants, but about 90% of the plant's chemical energy is lost in the process.
  2. The second law of thermodynamics helps to explain why large carnivores—lions and killer whales, for example—are relatively rare.
- **D.** The second law of thermodynamics is often invoked by creationists to prove that life could not have evolved from nonlife.
  1. Living things are exceptionally ordered states. Such an unimaginably ordered system could not possibly arise spontaneously.
  2. Locally ordered states arise all the time. Every time a plant grows, new highly ordered cells are formed.
- **E.** One question that science has not yet fully addressed is the tendency for highly ordered complex states, such as life, to arise locally. Perhaps, someday, there will be another law of thermodynamics.

Essential Reading:
Trefil and Hazen, *The Sciences: An Integrated Approach*, Chapter 4.

Supplemental Reading:
Morris, *Time's Arrow*.

von Baeyer, *Maxwell's Demon*, Chapters 5–17.

Questions to Consider:

1. Is it impossible for a disordered system to become ordered spontaneously? Would such an event violate the second law of thermodynamics?

2. The number of different arrangements of 52 cards in a deck is approximately 8 followed by 67 zeros! Each of these possible arrangements is a unique sequence, so why do we think of certain arrangements as more ordered than others? (Hint: For any given sequence, how much information is required to describe the arrangement of the 52 cards?)

Lecture Ten—Transcript
Entropy

Every natural system uses energy. Energy is one of the handfuls of unifying themes in all science. Indeed, every natural system has to obey the two laws of thermodynamics. The first law says that energy can change form from one type to another, to another, as many times as you want, and the total amount of energy is constant—that's the good news. But there's a downside, and that's the second law, that there are restrictions on how energy can transfer. Heat must flow from hot to cold, spreading out, evening out. There's a built-in direction to the way we use energy.

Yet this description of energy fails to capture the depth and the subtlety of energy in its role in our lives. As scientists of the 19th century thought about the implications of energy, they eventually came to a startling realization—the great principle that every isolated system becomes more disordered with time. In this lecture, the tenth of our series, I'm going to introduce the concept of entropy—the tendency of systems to become messier quite spontaneously, in spite of everything we do.

I'll begin by introducing the subtle idea of entropy, which follows logically from other statements of the second law of thermodynamics, the statements, for example, that heat spreads out, or that no system can operate with 100 percent efficiency. We're going to find out in this context that the second law holds tremendous power and sway over our lives. At the core of the concept of entropy are such human experiences as aging and death, decay. The second law defines the very direction of time.

I'm going to end the lecture by examining some of the far-reaching aspects of entropy, including the so-called "heat-death" of the universe, and the direction of time. There was a recent commercial for auto insurance on television, and it began with the totally smashed car in the middle of a flower shop, destruction all around, piles of debris, broken glass, flowers strewn everywhere. Suddenly the car lurches backwards through the store's plate glass window, which reassembles miraculously. The horribly damaged car rolls up the street hitting various objects, which are put back into position, and gradually pieces of the car attach back on, and at the end of the commercial the car is in perfect shape in a parking place. That sequence didn't violate the first law of thermodynamics at all, but we know that something is horribly wrong with that picture. We laugh because it defies physical logic.

The discussion of the second law of thermodynamics thus far is focused on the behavior of heat energy, and one statement of the second law is that heat flows spontaneously from hot to cold. Another equivalent statement is that you can't build an engine of 100 percent efficiency that converts heat to work. But as useful as these ideas may be, they miss the real essence of the second law. There are far-reaching consequences of the second law that go beyond just the way heat energy behaves. In its most general form, the second law of thermodynamics comments on the state of order of the entire universe.

I'm sure you've noticed the tendency of things in your everyday life to get messier. A new briefcase, you buy a briefcase, you buy a coat, or a new car, and it's spotless, it's shiny, it looks beautiful. A year later you look carefully, and the briefcase has scuff marks on it, the coat's a little bit frayed, the car has dents and dings and is going to be dirty. You can clean your home from top to bottom, and a week later, it's messy. It's got dust; things are strewn around. Our bodies too—they experience this inevitable aging, the decay. It seems like it's an irreversible process.

All systems in the universe have this tendency to become more disordered with time. A week ago I bought my wife a bunch of flowers, and they were brand new, they were beautiful. They're looking pretty sad right now, and the flowers, after a week they've become something that's only worthy of throwing into the trash. Of course, thanks to The Teaching Company, these are now a tax deduction, so I guess it's not all bad news.

You can devise all sorts of everyday experiments to show this type of behavior, and I just want to share a few of these with you. For example, you've all had the experience of shuffling a deck of cards, and you could order those cards into a perfect array, but as soon as you shuffle them, they become less ordered. And now they're not in as good order. Now, you could keep shuffling these cards over and over and over again, a million times, and they would never come back to the original ordered state. They would never be as ordered as they were when we started.

You can take a jar and put colored candies in them—two different types. And you can layer those, but if I shake the jar around just a little bit, all of a sudden they're no longer ordered. Now I could shake this jar as many times as I wanted to, and they'd keep coming up with new ordered arrangements, but they'd never be as perfectly ordered as they were when I started this.

Yet another example. You can take a jar of water and put a drop of food coloring in. To begin with, I had a highly ordered state with perfect food coloring and a clear glass of water, and as things go on, you know what's going to happen—that food coloring is going to spread out, diffuse out, throughout the entire jar. These examples are precisely analogous to a situation where a hotter and a colder object come together. The heat energy of the hotter material gradually diffuses into the colder object.

In every one of these cases, we could, if we wanted to, reverse the process. I could easily reorder the cards just by spreading them out and putting them in order. I could boil the water and recover that drop of food coloring. I could fix the candies—I could put them back in a layered order if I wanted to. And in theory, if we knew more about human aging, we could reverse that process too. There's no intrinsic reason why humans could not extend lifespan many, many fold if we knew how to fix the gradual damage, the gradual aging process.

All it takes in theory in all of these cases is the use of energy. You take energy, you can reorder a system that has become disordered. The concept of entropy, the disorder of a system, was introduced in 1865 by a German physicist, Rudolf Julius Emmanuel Clausius. Clausius was born in Pomerania, which is now a part of Poland, in 1822. He studied at several German universities; he graduated with distinction from the University at Halle. He accepted a professorship at the Imperial School of Artillery and Engineering in Berlin, and subsequently he taught in universities in Zurich and in Würzburg and Bonn, where he died in 1888.

Clausius devised the concept of entropy in order to quantify this natural tendency of systems to become more disordered with time. Clausius synthesized the ideas of Carnot and Joule and others, and he published the first clear statement of the two laws of thermodynamics in 1850. It was the first time in a publication there was a statement about these two laws that we've been talking about that gradually came to being understood over a period of many decades in the 19th century.

The second law in this original statement that wasn't presented in any rigorous mathematical form, Clausius just noted the tendency of systems to become more disordered with time. He realized that for the second law to be quantitatively useful it demanded a new, rather abstract physical variable. This variable he called entropy—which is a word derived from the Greek word for transformation. He defined entropy purely in terms of heat and temperature. Entropy, in fact, is the ratio of heat energy over temperature. This definition has very important applications in chemistry and engineering, where heat is used to do work. But for most of us, and I think this is including most scientists, this definition—heat divided by temperature—is a strange, very non-intuitive concept.

What Clausius did was to study steam engines. He observed the behavior of steam engines and realized that this ratio of heat divided by temperature must either remain constant or it has to increase. It never decreases. Stated another way, in any heat engine, the heat energy divided by the temperature of the cold reservoir is always greater or equal than the heat energy divided by the temperature of the hot reservoir. Thus, the entropy of a closed system cannot decrease. Entropy can be constant or it can increase. And in most instances, because there is always some natural escape of heat energy into the surroundings, in most cases entropy increases.

Every physical substance has a quantity of entropy just like it has a quantity of energy. You can measure it and give it a unit. At least in principle you can measure the entropy of a system by measuring its temperature and its total heat energy, although it turns out that that's not so trivial a thing to do. So in this case, you can summarize the two laws of thermodynamics in a very short way, as energy is constant, entropy tends to increase. And that's the shortest way of saying the first and second laws of thermodynamics.

I confess that all of this is very abstract stuff. You can express everything I've just told you in mathematical form with equations. And lots of scientists and lots of engineers have learned to use those equations to do all sorts of very important things with heat and energy and so forth. But my impression is that the number of scientists and engineers who really have a deep intuitive understanding of entropy is quite small.

This is a really abstract concept, this concept of entropy. So I want to help you with this, with maybe a more intuitive way of thinking about entropy. Why should things become more disordered? You can do this by thinking about the natural world at the scale of atoms. I haven't really introduced atoms yet, but all matter, all things around us, are made of atoms—these are tiny particles, they come together to form chemical bonds, to form the various solids, liquids, and gas structures we see around us. And you can think about entropy in terms of the properties, particularly the energy, the kinetic energy, of individual atoms. Heat spreads out because some atoms with more kinetic energy collide with atoms that have less kinetic energy—the slower atoms, if you will. And so, eventually the kinetic energy of atoms averages out and this is why heat spreads out; it's just a physical process of vibrating atoms that keep colliding into each other, whether they're in a gas state, or a liquid state, or in a solid state. Concentrated heat thus spreads out.

Ultimately, the order of any system can be measured by the orderly arrangement of these smallest parts: these atoms. Think about some of the materials that you see around you. A crystal of table salt has a highly ordered arrangement of atoms; you have sodium and chlorine atoms, alternating sodium-chlorine, sodium-chlorine, sodium-chlorine in a very orderly arrangement. A lump of coal also has a highly ordered arrangement of atoms that contains a lot of this ordered energy with carbon-carbon bonds, and bonds between carbon and hydrogen particularly—those are very energetic bonds. You can dissolve the salt in water. You can burn the coal, and you disrupt those bonds. The material becomes more disordered. The entropy increases as you release that heat energy.

The arrangement of atoms doesn't have to refer specifically to a solid. You can talk about the arrangement of atoms in a gas as well. And you can imagine gas atoms having different temperatures. They have different velocities. Imagine what happens if you have two reservoirs of gas that are at different temperatures and suddenly you mix them. You dump two bottles of gas—one is very hot and one is cold, for example. The separation of the gas into two separate populations is a more ordered state. It's like having all the red marbles in one part of the jar and all the blue marbles in another part of the jar. And as soon as you mix them together, the hot atoms in the hot gas start colliding with the colder atoms and the temperature averages out, because the average velocity of the gas atoms averages out. The entropy, therefore, the disorder of the system, increases.

This definition of entropy, as a measure of the degree of disorder, may seem fuzzy, but it was placed on a very firm quantitative footing in the late-19th century by the Austrian physicist, Ludwig Boltzmann. Let me tell you about Boltzmann. He was born in Vienna in 1844. He studied at the University of Vienna, where he spent most of his professional life as a professor of theoretical physics. He was said to be an imposing man of great physical strength, combined with a kind of sensitivity and humor. He also suffered from severe bouts of depression, and it was during one of these spells that he took his own life in 1906.

Boltzmann used what's called probability theory to demonstrate that for any given configuration of atoms, the mathematical value of entropy is related to the number of different possible ways you can achieve a particular configuration, and I'll show you what I mean. Entropy, in fact, is the logarithm of the number of configurations, that's the mathematical form. This is a subtle and a very profound point.

Let me try to illustrate what Boltzmann meant. I want you to consider, rather than the millions and millions of atoms that form even a tiny part of air, let's just look at six balls numbered one through six. Three of them are yellow, three of them are orange. And think about all the different ways to arrange these balls. It turns out that there's $1 \times 2 \times 3 \times 4 \times 5 \times 6$ different arrangements, a total of 720 different ways just to line up six balls. And if you go through the math, it turns out in exactly 36 of those ways, you can rearrange the balls so you still have an ordered arrangement of three yellow followed by three orange. And you can imagine all different ways of moving these around—36 different ways of doing that.

But that's 36 out of 720 different ways, so if I randomly threw down all six balls in a row, you'd only have one chance in 20 of having three yellow followed by three orange. All the other arrangements would be different. A lot of them would have things mixed up—when the oranges and the yellows are mixed up in various ways—and most of the arrangements would then appear more disordered, and that's a situation with only six different balls.

Now imagine if I had a million balls, or a trillion balls, or as in atoms, a trillion, trillion, trillion atoms. You could imagine there are a few ordered states in which all the hot atoms are in one side of the room and all the cold atoms are in the other side of the room. But the chances, the probability that you'd achieve such a situation are so infinitesimally small compared to all the disordered arrangements that the disordered arrangements always dominate in the natural world. It's just a matter of probability. So this is the key to understanding entropy. It is the microscopic manifestation of probabilities. It's just like going to Las Vegas. What are your chances of winning? Well, you're talking about ordered states of the natural world, your chances of winning are very small indeed.

I'm going to shift gears now and look at some of the consequences of the second law of thermodynamics. I'm going to tell you about the so-called "heat death" of the universe, the nature of time, and also the central biological concept of trophic levels. These are three of the consequences of the second law.

Arthur Eddington was a very influential 20th-century British cosmologist, about whom we're going to learn a lot more in Lecture Thirty-Two. Well, he said, and I quote, "the law that entropy always increases, the second law of thermodynamics, holds, I think, the supreme position among the laws of nature." Well, why did he say this? Why did he think it had such deep and far-reaching consequences?

One of the more sobering consequences of the second law is the idea that the heat in the universe must eventually spread out. Life on Earth is possible because the Sun provides a steady stream of heat, a radiant energy to the Earth's surface. The Earth also holds a vast reservoir of internal heat. But if you think about it, in the 1890s people began to realize that that heat in the Earth is still radiating out into the coldness of space. The Sun's energy is radiating out into the coldness of space. Eventually, perhaps billions of years down the road that heat energy is going to disperse and distribute evenly. The ultimate fate of the universe, therefore, must be for the heat to spread out uniformly, as in the second law.

Now we don't necessarily have to worry about this, especially with the discovery of nuclear energy, which provided vast amounts of energy still preserved in the universe today, indicates that it's going to be hundreds of billions of years or more before this heat death really takes place. But nevertheless, it's a sobering consequence to think that the universe itself may end in this kind of whimper-like state, with everything in a uniform low temperature, no life possible, nothing of interest ever going on. Indeed, no energy transfer would be possible, because all the energy is taken up in matter at the exact same temperature. You can't do any work if you don't have two heat reservoirs at different temperatures.

The second law also tells us something about time, the nature of time. Throughout this course I refer to time in lots of different contexts, I talk about distance equals velocity times time, power equals force divided by time. We experience time as one of the major variables in our lives. We try to deduce historical aspects of things and time becomes a very important parameter in those kinds of discussions. But what exactly is time? Talk about an intangible physical quantity. What is the nature of this quantity, time? Why does it seem to have a direction from past to future? These are really deep questions and in many senses, these questions are more philosophical, they are beyond modern science, but the second law of thermodynamics gives us some tantalizing insights.

Again, I want you to imagine playing your favorite movie backwards, imagine playing a videotape backwards. Many scenes are going to appear perfectly normal, drifting clouds on a windy day would look very normal, blowing leaves, even a gently flowing stream (if you couldn't see a very steep slope) would look perfectly natural either way, but some forms of motion would look very, very strange: people diving out of a swimming pool, unbreaking an egg, unshooting an arrow. These are impossible in the real world. You have to ask yourself why are some actions irreversible? And I think it all ties with this tendency of entropy to increase. The tendency of entropy to increase rather than decrease defines the arrow of time. It defines the direction in which time moves in our everyday experience. When you shake a jar it becomes more disordered. When you shuffle a deck of cards it becomes more disordered. When you put food coloring in a jar of water, see? It spreads out quite evenly. That's the direction of time, that's the arrow of time. Glass shatters, milk spills, organisms die and decay—the arrow of time.

Of all the great natural laws, the laws of motion, the laws of gravity, the first law of thermodynamics, electricity and magnetism, as we'll see, only the second law of thermodynamics incorporates this arrow of time. Somehow the second law defines the direction of time. Whether that's a cause or effect, whether we're merely describing some aspect of the natural world that is, it's still a fascinating idea, and that's why scientists who think about energy also think about time.

Let's consider yet another example of this second law and its consequences. All living things are subject to the limitations of the laws of thermodynamics. That's pretty obvious. We're living systems, and we're natural systems. We see this in many, many different ways, including the pervasive biological concept of trophic levels. Every organism has to compete for a limited supply of energy. Plants are in this first trophic level because they get their energy directly from the radiant energy of the Sun. In the second trophic level, we find herbivores, these are animals that eat plants. But almost 90 percent of the energy that's contained in plants' chemical substance can't be used—there's inefficiency here. So, while plants get about 10 percent of the Sun's energy, herbivores only get about 10 percent of a plant's energy. And then come the carnivores in the third trophic level, and carnivores only get about 10 percent of the energy contained in the herbivores, in the meat that they eat.

So you see the second law of thermodynamics coming in and the efficiency of living systems. Plants are 10 percent efficient, herbivores are 10 percent efficient; carnivores are 10 percent efficient. So as you go up the food chain to the largest carnivores, this explains why there are so very few large carnivores compared to the number of herbivores, and then compared to the total biomass of all the different plants, as a natural consequence of the second law because you're passing energy through a system, but it's not an efficient passing of energy. It takes 10 herbivores for every carnivore. And if you have a large carnivore that eats other carnivores, those are really going to be rare.

This concept of trophic levels, by the way, has real important consequences for the world food management. It takes many times more energy to produce meat and poultry than it does to produce a certain amount of grain, with similar nutritional value, with a similar number of calories. So for that reason, as the world becomes more and more crowded and more populous, we're going to have to look more and more to plants as our most efficient source of food energy.

The second law of thermodynamics, it turns out, is one of the most misunderstood of all natural laws. The law is often invoked by creationists to prove that life could not possibly have evolved on Earth. Now let me tell you about their rationale. It goes something like this. They say living things are exceptionally ordered states. Every single cell in our bodies—and we have a hundred trillion cells in our body—every cell has countless trillions of trillions of atoms that are precisely and intricately ordered. Certain atoms, if you just move certain atoms out of position, the cell doesn't work anymore. If those cells don't work anymore, the body doesn't work anymore. So life is probably the most ordered system we can imagine. The human brain is probably the most ordered and intricate object we can conceive of. Well, given the second law, it seems since systems tend to become more disordered with time, it's inconceivable that such an ordered system could have evolved within the context of the second law. Because systems tend to become more disordered, not ordered like the human brain, so how could you evolve life? For this reason, the creationists say you have to have a miraculous creation of life. You can't imagine a natural system evolving life in this way.

This argument, though, misses an important aspect of the second law. You can have locally ordered states that arise, and they arise around us all the time. Think about a pan of salt water put out in the Sun. The Sun evaporates the water and salt crystals, incredibly ordered salt crystals form, where before there was a very disordered arrangement of sodium and chlorine ions in the water. The key is that the Sun adds energy to the system locally. But if you think of the entire system, including the Sun and the Earth, the Sun is becoming more disordered. It's burning up its fuel, and as it burns up its fuel, the total system is becoming more disordered, but locally, you can have pockets of increasing order.

Every time a plant grows, every time you get a cut and it heals, in that sense you are sort of overcoming the second law, because you're taking energy and you're fixing damage, you're making things better when before they were worse. So we see every spring as new plants arise all around us, we're seeing these so-called violations of the second law, but they're not violations at all, it's just a reshuffling of energy, it's systems that have learned how to use energy and collect energy from their surroundings. Their surroundings have entropy that's increasing, but the individual objects then become more ordered, and that's how you can explain the origin of life; it's just another example of this very common, natural process that occurs around us, albeit, a remarkable one. Let's not get away from that.

You might think that energy is, by now, so well studied that we know it all. We know pretty much everything there is to know about energy, but I think that may not be so. There's one nagging aspect of energy that I believe science has not yet fully addressed. I may be a minority on this, but think about it, there's a tendency of physical systems to spontaneously become complex locally, as I just said, even though they become globally disordered. What do I mean by that? Well, there's a seemingly universal tendency for systems with large numbers of components to display complex behavior that you wouldn't normally see in the individual components. And you can see so many examples. Individual sand grains can just lie there, but if you put enough sand grains together, you get dunes, you get ripple marks, you get all sorts of complex behavior that you'd never predict from a single sand grain.

Atoms form planets, and then life forms from those planets. And then consciousness forms from life, and you have these emergent, complex processes that seem to arise spontaneously out of local order in an otherwise large system in which the entropy is increasing.

In each of these cases, you see local states of extremely high order arising from a large collection of small particles, whether they be cells or atoms, or so forth. These are what we call emergent properties. Complexity seems to just arise in the universe, and as yet there's no law of thermodynamics that describes how complexity can arise. It doesn't fit into the first law. It doesn't fit into the second law. Perhaps there's a third law. Part of the answer may lie in deep connections between the concept of entropy and the concept of information. But we're really a long way from a full understanding of this point. In any case, I'm willing to bet personally, that someday, perhaps soon, there's going to be a third law of thermodynamics. That's one of the exciting things about science—we keep coming up with new ideas. Science is an endless frontier.

Let me summarize this lecture on the second law of thermodynamics. This law provides the framework for understanding so many of the directional events in our lives. The simplest statement of the second law is that heat tends to spread out evenly. And that's the special case of this more general tendency of systems to become disordered, a tendency that is quantified by the concept of entropy.

Entropy can be defined formally as the ratio of heat energy divided by temperature, or it can be treated statistically as a measure of the number of possible configurations of a system—that is, the probability of having an ordered versus a disordered state. In either case, the entropy of a closed system will not decrease spontaneously. And then by defining the sequence of possible events, the second law of thermodynamics provides a definition of the arrow of time.

Energy is going to return over and over and over again in these lectures as an underlying theme. But first, we have to take a closer look at two of the forces that play a vital role in our everyday life—that's electricity and magnetism—and that's the subject of the next lecture.

Lecture Eleven
Magnetism and Static Electricity

Principle: *"Magnetism and static electricity are forces that can be either attractive or repulsive."*

Scope:

Newton's laws of motion defined the concept of forces and established a logical protocol for their study. Magnetism, the force that causes one end of an iron needle to accelerate in the direction of the Earth's North Pole, was of great importance in an age of ocean exploration and commerce. Every magnet has two poles, designated north and south. When two magnets are brought together, opposite poles attract and like poles repel. English physician and physicist William Gilbert (1544–1603) showed that the Earth, itself, is a giant magnet with its own field and that smaller magnets, such as compass needles, align themselves in this field.

Static electricity, a subtle and pervasive force, was little more than a curiosity in the 18th and early 19th centuries. Amber, glass, and a variety of other materials become electrically "charged" when rubbed with fur or silk. Electrically charged objects have either too many electrons (a negative charge) or too few electrons (a positive charge); like charges repel and opposite charges attract. Charles Coulomb (1736–1806) determined the force law between two charged objects: force equals the product of the two charges divided by the square of the distance between them, times an appropriate constant—a relationship similar to the equation for gravity. The law is different in that the electrostatic force may be either attractive or repulsive, and the electrostatic force is much stronger than gravity.

Outline

I. Newton's laws of motion define the concept of a force—a phenomenon that causes mass to accelerate. While Newton applied his laws only to the force of gravity, they established a logical approach to the identification and characterization of other natural forces. One of the most important and mysterious forces in Newton's day was magnetism, which causes one end of an iron needle to accelerate in the direction of the Earth's North Pole.

A. The study of magnetism was important in an age of ocean exploration and commerce.
   1. Magnetic rocks were discovered by the ancient Greeks in a region of Asia Minor called "Magnesia." These rocks attract pieces of iron, and they attract or repel each other.
   2. Ancient scholars knew that a magnet suspended by a string will pivot so that one end (the north pole) points north, and the other end points south. A compass is a magnetic needle on a pivot.
   3. When two magnets are brought together, opposite poles attract and like poles repel.
   4. A piece of unmagnetized iron can be magnetized by repeated stroking of a magnet.

B. Compass makers, such as British sailor and instrument maker Robert Norman (c.1550–1600), attempted to improve the sea-going compass. His principal work, *The Newe Attractive* (1581), described the tendency of compass needles to dip.
   1. Dip was a nuisance to compass makers, because it was difficult to get their needles to balance properly.
   2. To study the effect of dip, Norman pivoted a compass needle on a horizontal axis and thus established the angle of dip.
   3. Norman devised an ingenious experiment to test whether there is a net force on a magnetic needle.

C. Various ideas about the behavior of magnets were synthesized in the research of English physician and physicist William Gilbert (1544–1603).
   1. In *De Magnete* (1600), Gilbert demonstrated that every magnet has two poles; broken fragments of a magnet are themselves complete magnets.
   2. Gilbert proposed that the Earth, itself, is a giant magnet with its own field and that smaller magnets, such as compass needles, align themselves in this field.
   3. This kind of field, with a north and south pole, is called a dipole. The Earth, the Sun, and many other bodies are large dipole magnets.

4. Some volcanic rocks "freeze in" the orientation of the Earth's magnetic field when they cool.
5. British experimentalist, Michael Faraday (1791–1867), performed the simple demonstration of sprinkling iron filings near a magnet.

II. Static electricity is a subtle, yet pervasive, force. In the 18th and early 19th centuries, static electricity was a curiosity, of little practical concern, but still worthy of investigation.
   A. William Gilbert, who studied the Earth's magnetic field, also studied electrostatics. He found that some substances become electrically "charged" when rubbed with fur or silk—amber, glass, and some minerals, for example.
      1. Some materials, called insulators, hold this charge.
      2. Conductors, especially metals, do not hold their own charge, and they drain charge away from insulators.
      3. Charges can be passed from one object to another by touching.
      4. Electric friction machines, in which a belt of rubber, cloth, or fur is rapidly spun over a piece of amber or other insulating solid, are able to develop large charges. Such machines were used well into the 20th century to produce extremely high voltages in physics experiments.
   B. Electrically charged objects exert forces on each other—forces that can be measured systematically by suspending balls of styrofoam on threads. Two electrically charged objects can either attract or repel each other, similar to a magnet.
      1. Benjamin Franklin (1706–1790), American statesman and signer of both the Declaration of Independence and Constitution, devised an explanation for these curious observations. Franklin suggested that electric charge is a fluid (what we now know as electrons) and that all uncharged substances have some fixed amount of this fluid. A material becomes charged when friction adds or removes charges. If it has too many electrons, it has a negative charge; if too few electrons, it has a positive charge.
      2. The behavior of these balls can be explained if like charges repel and opposite charges attract. The behavior

of Franklin's electrical fluid modeled all observed electrostatic behavior long before the discovery of electrons.
3. Franklin applied his knowledge in his invention of the metal lightning rod, which conducts electrical fluid harmlessly into the ground.

C. Charles Coulomb (1736–1806) conducted meticulous experiments on the electrostatic force and determined the exact force law between two charged objects: force equals the product of the two charges divided by the square of the distance between them, times an appropriate constant.
1. This equation looks very much like the equation for gravity.
2. The law for electrostatic force is different from gravity in two important ways: the electrostatic force may be either attractive or repulsive, whereas gravity is always attractive, and the electrostatic force is vastly greater than gravity.

D. Static electricity is not a central topic of physical research these days, but we do have to deal with its consequences.
1. Lightning occurs when violently agitated raindrops in a cloud pick up an electric charge through friction.
2. A xerography machine uses static electricity to apply black plastic powder to paper.

Essential Reading:
Trefil and Hazen, *The Sciences: An Integrated Approach*, Chapter 5.

Supplemental Reading:
Harre, *Great Scientific Experiments*, Chapter 3.

Questions to Consider:
1. If the Earth had no magnetic field, how might the force of magnetism have been discovered? Would scientists have thought it worth studying?
2. Why are effects of static electricity so much more noticeable on cold, dry days?

Lecture Eleven—Transcript
Magnetism and Static Electricity

Newton's laws of motion define the concept of a force—that's a phenomenon that causes a mass to accelerate. While Newton only studied one of these forces, the universal force of gravity, he's established a protocol for looking at all the other forces in nature. In Lecture Eleven, I'm going to tell you about two of the most ubiquitous yet complex phenomena in nature in our daily lives—that is magnetism and static electricity. One of the great and surprising principles of science is that magnetism and static electricity are forces that can be either attractive of repulsive. My objectives in this lecture are to describe the historical developments of understanding the phenomenology of magnetism and electricity, which seem to be two very different forces.

For magnetism we're going to focus on natural magnetic materials, the Earth's magnetic field, and the behavior of compass needles. In the case of static electricity, we're going to look at the concept of an electric charge, and review the behavior of charged objects. In subsequent lectures then, we're going to find that electricity and magnetism, these seemingly unrelated phenomena, are indeed, closely related. They're intertwined. But in this lecture, that won't be quite clear yet because we're looking at this from the historical context, and in the early days people didn't recognize this connection.

Before beginning the exploration of electricity and magnetism, let's review where we've come in the first ten lectures of this series. In the first two lectures we examined science as a way of knowing, as a process of asking and answering questions using that scientific method of observation, pattern recognition, hypothesis, theory, and then prediction, which led to more observations. These are the four steps of the idealized cycle, although in many cases in the natural world scientists have to use intuition and make intuitive leaps of kinds that wouldn't be strictly dictated by that scientific method. The next two lectures explored how the scientific method was used by a whole succession of scientists to develop the modern view of the heliocentric, the sun-centered solar system. New observations of planetary positions led from the Ptolemaic Earth-centered view to the Copernican model with circular orbits, and then to the realization by Kepler that orbits are, in fact, elliptical. Isaac Newton and his universal laws of motion and gravity were the subjects of Lectures Five and Six, while the last four lectures have examined energy—the ideas of the first and the second laws of thermodynamics.

Throughout these lectures we've touched on the concepts of mass, of motions, of energy, of forces, and these are the central attributes of the physical universe. One of the most important and mysterious forces in Newton's day was magnetism, which causes one end of an iron needle to accelerate in the direction of the Earth's North Pole. The study of magnetism was, of course, of extreme importance 500 years ago, because it was an age of ocean exploration, and the compass was the key tool in preventing you from getting lost, especially in days when the Sun and the stars were not visible. So the compass, this simple device—a device which has a magnetically aligned needle which points north and points south—this simple object has played such a vital role in the discovery of the natural world.

Magnetic rocks, called magnetite or lodestone, were discovered by ancient Greeks. They were discovered in a region of Asia Minor called Magnesia. And so these rocks, these rocks that attract pieces of iron, where you actually can pick up paperclips and so forth with a rock, this naturally occurring material began the study of magnetism. Magnetite is an abundant ore of iron that's found all over the world and so it was widely available to the ancients, and the phenomenon of magnetism therefore was understood and observed quite early on. A few basic features of magnets were well known to the ancients. A magnet suspended by a string, for example, can pivot, and if you have a second magnet you can apply forces to that. You can see the forces acting as the magnet swings around under the influence of a magnetic field. The compass then is merely a magnet that's suspended on a pivot point, and it can swing around in the Earth's magnetic field.

When two magnets are brought together, you find the opposite poles attract each other, called north and south poles, but like poles—two north poles or two south poles—will repel each other. You've probably had that experience if you've ever played with magnets. It's a great thing to do with your kids, to get a set of magnets and see how they behave. A piece of unmagnetized iron, like a nail, can be magnetized simply by stroking it with a magnet. This is something else you can do. You can take a powerful magnet and take a nail and just run the magnet along it a few times, and that nail then, if you do it enough times, will become a magnet even though it was not magnetized before.

Perhaps the first researchers to conduct extensive studies on magnets were the compass makers themselves. And you can imagine why that's true. If you're going to build compasses, you have to understand how magnetism works. You have to understand how to magnetize needles and so forth.

One such craftsman was Robert Norman, who lived from about 1550 to 1600. Records on these early instrument makers aren't so precise as those for scientists and people who published things. Norman was an instrument maker, and so as a tradesman, his records are not so well preserved, so very little is known about him. We know that he was, for many years, a British sailor, and that he became a maker of navigational instruments following that career as a sailor.

Rather than simply duplicating the design of others, however, Norman attempted to improve the seagoing compass. He then wrote some of his experiences in a book called *The Newe Attractive*, published in 1581. And this described the curious tendency of a compass needle to dip. The dip is somewhat of a nuisance to compass makers. This is the tendency of a compass needle not only to align itself north and south, but also to adopt an angle dipping into the Earth. Norman describes an experience trying to make a compass needle and the frustration. This is why he's trying to correct a particularly beautiful and elaborate, decorated needle. You can imagine these iron needles with all sorts of engravings and beautifully shaped, and so forth. He created such a needle for a very elaborate compass, and the trouble is it kept dipping down touching the bottom of his compass, and so it didn't work very well.

Here's what he said: "I cut it too short, and so spoiled the needle wherein I had taken such pains. Hereby being stricken into some choler, I applied myself to seek further into this effect." Well, he didn't just get angry, he got even. He began to study the dip, or the declination of the magnetic needle. What he did was a very clever idea. He took the needle and he put it through a cork, so it was precisely balanced. He then had it floating in water, and the needle then would orient itself in water, it would actually orient itself up and down, so he was able to measure the magnetic needle by exactly balancing the weight of the needle in water.

Norman concluded by suggesting that the compass needle acquires a virtue in spherical form extending about the stone. The idea of this virtue extending about the stone was the first real suggestion that there was such a thing as a magnetic field—imaginary lines extending out from the magnetic object, which then interacted with the Earth's magnetic field. He also saw a very practical side to this magnetic dip that he was studying. He realized that since the dip varies with latitude on Earth, it might provide an extra navigational aid. You could actually use the dip of the compass needle to help locate where you are, not just north and south, but also how far north or south you are.

Various ideas about magnets were synthesized in the research of the English physician and physicist, William Gilbert, who lived from 1544 to 1603. Gilbert must have led an amazing life. He graduated with a medical degree from St. John's College in Cambridge. That was in 1569. That's about the same time that Tycho Brahe was making his first observations of the heavens—the astronomical observations. And at that time, Galileo was just 5 years old. So this was a very exciting time in the history of science. He began practicing medicine in London in the mid-1570s, and by 1581 he was one of the most prominent physicians in the city, rising to the presidency of the College of Physicians in London, a very distinguished position indeed. In 1600, he was appointed personal physician to Queen Elizabeth I. He served in that capacity until her death in March of 1603, and William Gilbert died just a few months later of the plague, in 1603.

His great scientific work is *De Magnete*, published in 1600. Gilbert catalogues the properties of magnets. He demonstrated, for example, that every magnet has two poles. He also showed that broken magnets—you can take a magnet and break it again and again and again, and every fragment of the magnet also has a north and south pole. Gilbert proposed that the Earth itself is a giant magnet with its own field, with its own north and south poles. To test this idea, Gilbert constructed a spear of magnetite, of that natural magnetic material, and documented the nature of the magnetic forces around this spherical object. This kind of field with a north pole and a south pole is called a dipole field—a two-pole field. The Earth, the Sun, many, many other bodies in the heavens have this kind of dipole magnetic field, and so does any magnet. Any magnet has a dipole field—a north and a south pole. So that's called a dipole.

The Earth's magnetic poles don't correspond exactly to the rotational north and South Pole. The Earth rotates on an axis, and that defines the North Pole and the South Pole. The magnetic pole, however, is several degrees off and it wanders slightly, it moves around from year to year. Furthermore, over long periods of geological time, perhaps once every half million years or so, the Earth's magnetic field has this remarkable property of actually flipping. The Earth's north pole ends up near the south, and the south pole ends up near the north. This flip occurs, perhaps on average once every half million years. We'll talk more about that when we talk about plate tectonics later on in this course.

The idea of the Earth's magnetic field flipping and moving around is very important in the field of geology. Some volcanic rocks freeze in the magnetic orientation of the Earth when they cool because volcanic rocks have little magnetic minerals—in fact, magnetite—that act as compass needles. When the magma, the volcanic rock, is liquid, those compass needles orient to the Earth's magnetic field, and then those orientations are frozen in, so the cold volcanic rock has its own magnetic field and tells you what the magnetic field of the Earth used to be many millions of years ago. And this information is used by geologists to understand the variation of the magnetic field over the millennia.

Norman and Gilbert had clearly introduced the idea of this magnetic field, but it wasn't until the early-19th century that a gifted British experimentalist, Michael Faraday, who lived from 1791 to 1867 (and we're going to hear a lot more about him later on) performed a simple demonstration of sprinkling iron filings on a magnet. And I'm sure you've seen these illustrations of iron filings forming beautiful patterns outlining the magnetic field lines. That was a 19th-century discovery by Faraday who liked to do public lectures in science.

You see complexly curving field lines. They stand out starkly in this sort of experiment, and they're not simple lines radiating out from a point, as we talked about in the gravitational field—the idea of lines just moving straight out, but rather the field lines curve around from pole to pole. Along that axis, force varies over distance squared, just like electricity and magnetism, but they're curving field lines. Along other directions the force exerted can be extremely complex. The force is a function of the size and shape of the magnet as well as the position of the object (for example, a piece of iron within that dipole field), so the exact description of the magnetic force with a magnet, especially a curving magnet like a horseshoe magnet, can be quite complex, indeed, much more difficult to describe and model mathematically than a force field coming out of a gravitational object like the Earth.

Many mysteries about magnetism remained in Michael Faraday's time. One persistent question is whether it's possible to have a magnetic monopole, that is, a single north pole or a single south pole in an object, and that question is still very much with us today. Another very big question for Faraday and his contemporaries was what exactly caused a magnetic field in the first place? And to answer that question we have to wait until we learn a little bit more about electricity in the next lecture.

Okay, now I want to shift gears, and we're going to talk about the other great force, the other great phenomenon that people studied in these early years—it wasn't obviously related to magnetism in any way—and that's static electricity. This is a subtle, yet pervasive force in all our lives. You get static cling. You get frizzy hair on a dry winter day. Lightning is static electricity. The little sparks you feel when you shuffle across the room on a cold day and you touch a doorknob, that's static electricity.

In the eighteenth and early-19th centuries, static electricity was a curiosity. It was a natural phenomenon of little practical concern, but perhaps still worthy of philosophical investigation, because Isaac Newton had said any time something happens, there's a force involved, and clearly something was happening with static electricity. One of the people who studied static electricity in the early period was William Gilbert, who did so much to improve our understanding of magnets. He found that some substances become electrically charged when they're rubbed with fur or with silk. Amber, glass, some minerals, for example, also have this process. More rubbing produced a greater charge.

And we can demonstrate this quite easily with a very simple illustration. You can take a couple of packing peanuts—those little styrofoam balls you get in the kinds of packing you receive in packages in the mail—you just wrap them with aluminum foil and suspend them with a string. Then just take a comb and run the comb through your hair a few times, picking up static charge, and you'll see that the comb then causes attraction. You see there's a force involved here—something's happening. This isn't gravity; this isn't magnetism. That's something quite different, and so you can study this force.

It turns out that some materials, which are called insulators, hold this electric charge, whatever that might be. Other materials, especially metals, don't hold their own charge and the charge drains away from these conductors. Those substances are like metals and wires and so forth. You notice that in this particular arrangement we have insulating materials with a suspended what are called pithballs. You have a string that hangs down, but a little piece of conducting material at the end, which is insulated from its surroundings. It turns out that charges can be passed from one object to another just by touching. You saw that with a comb, combing through your hair. You touch one of these pithballs, as they're called, and you see the charge passing from one to another. You can transfer a charge by rubbing a piece of fur or glass of various combinations, and you can also get this transfer of charge from one object to another.

One of the most efficient and effective ways of doing this is something called an electric friction machine. In these machines you have a belt of rubber or cloth or fur that's spun rapidly against another substance, and as that spinning occurs you strip off a great deal of charge and develop very high electrostatic charges. Such machines were used well into the 20th century. Things called Vandegraaff generators were used to develop extremely high voltages for high-energy physics experiments, so these weren't mere curiosities in the 20th century; they were the best way to develop very high voltages. Electric friction machines, however, were first used for parlor games and novelties. A great novelty in the 18th century was called the electric kiss. What you'd do is you'd have a group of people standing around this friction machine holding hands, and a young engaged couple would be at opposite ends of the people, and as you'd crank up the electric friction machine, you'd develop a very high voltage, the two people would kiss and a spark would pass between their lips and everybody would jump.

Let me read you a historical account by the American historian, Bern Dibner. This is another example of using an electric friction machine. "An electrical discharge was sent through a company of 180 soldiers of the Royal Guard before the king of France. This occurred in the courtyard of the palace of Versailles, and the shock sent all the men leaping into the air at the same time, much to the glee of the king and his party. So dramatic was the demonstration that the king had it repeated in Paris using 700 Carthusian monks, who all jumped into the air together when jolted by an electric shock."

You can actually use one of these electric friction machines. I've brought one with me today, and will subject myself to its terrors. You merely turn the intensity up, and you see a belt moving here, and as I lean over…Oh!! You can get quite substantial shocks out of this, and perhaps my hair is also standing in the air. So I'm developing quite a voltage there. And we'll leave it at that. So this is a way of generating very high electrostatic charge, by rubbing a belt against another machine. That's just an electric friction machine that's been around for hundreds of years.

Electrically charged objects exert forces on each other. You can study these forces in a systematic way. And once again, the way that people did this to really study this systematically was to suspend various objects that could develop a charge from strings and so forth. And they could do this very systematically. I'll show you again. Look very carefully at what happens. You charge up the comb and at first you get it an attractive force. See? But then once the object gets touched after that attractive force, then it seems to be repelled, and so we have a very odd situation here of a force, that is, that once it can be attractive and it can be repulsive. So the electrostatic force differs from gravity in that way—that it can be both attractive and repulsive.

What's happening here? Turns out that Benjamin Franklin, the great American statesman, the great ambassador to France, the person who played such an important role in drafting the Declaration of Independence, signing the Constitution and so forth—he lived from 1706 to 1790—was Benjamin Franklin who was best known in his own day as a scientist, as a physicist who studied this electrostatic phenomenon. He was called one of the leading electricians of his day. That was the title they gave to people who studied electrostatic phenomena.

Let me tell you about Ben Franklin. He's an amazing character. He was born in Boston in 1706. He received his early education in Massachusetts, but is best known for his activities in Philadelphia, which was at the time, believe it or not, the second largest city in the British Empire. Philadelphia was second only to London in size. Franklin founded the first library in America. He founded the first fire department in Philadelphia. He was extremely active in the city life there, in both the state of Pennsylvania and also, of course, in the American colonies as a whole. Franklin gained great fame as a successful inventor, notably of the Franklin stove, and then later of the lightning rod, about which I'll tell you. If you would ask educated Europeans about Benjamin Franklin in the 1760s and 1770s, most would have known of him and they would have known of him for his scientific work.

Franklin suggested that the electrostatic charge, the electric charge that we saw here, is a kind of fluid, is what we now call electrons, although Franklin did not know that and that was not discovered until a good century later. A material becomes charged when friction removes or adds this fluid from one object to another. So a normal object that you pick up, like a comb, is balanced and has a neutral amount of this fluid, exactly enough fluid, but if you comb it through your hair then Franklin wasn't sure one object or the other, either the comb or your hair is gaining more of this fluid, and the other's getting less of it. And so each become charged—the hair becomes charged, the comb becomes oppositely charged by the transfer of this one kind of fluid, which should be in balance, but in a charged object it's not.

The behavior of these balls then, can be understood by Franklin's idea. When you first charge an object, the comb is charged, the ball is not, but the fluid in the ball, that is the ball's electrons, are distorted by the comb's electric charge. At first, the ball is attracted to the comb, but then they touch and both objects are charged, then they repel each other. The initial attraction is because the ball that's uncharged has the fluid moving away from the charged comb so it gets pulled in, and then suddenly it has an excess, and it gets pushed apart. So, you have the attraction and repulsion both shown by these two objects.

Franklin was an eminently practical man. He recognized that certain materials take electric charge and transfer electric charge, and other materials were insulators. And he realized that you could apply this knowledge to protecting lives, to saving lives, to the invention of the lightning rod. The lightning rod is such a simple idea. Franklin realized that lightning was a form of electricity, of this electrically charged object. In fact, it's said that Franklin went out in a thunderstorm with a kite with a metal string to actually prove to himself that lightning had this kind of electrical property. I don't think Franklin did that experiment (although others did) and showed beyond a doubt that lightning was a form of static electricity.

So, if you could find a material that would safely conduct that fluid away from your house down into the ground, you would prevent houses and other buildings from being hit by lightning, and that's what the lightning rod did. It's just a simple piece of metal and metal wire, which goes from the roof of your house down into the ground, therefore conducting the electrical fluid away from the house. And it was this understanding of the electrical fluid that allowed Franklin to make this great invention.

You can quantify the forces that are involved in these electrostatic demonstrations. There is a force involved, and through very meticulous and careful experiments, you can measure the exact amount of force. You can measure the way it varies with distance and with the amount of charge, and so forth.

And that was left to the job of Charles Coulomb, who lived from 1736 to 1806—a great French physicist, who conducted meticulous experiments on electrostatic forces, and he determined the exact force law between two charged objects. Force equals the product of two charges divided by the square of the distance between them. The equation: force equals the constant charge of the first object times the charge of the second object divided by distance squared.

Does this equation look familiar to you? Does it remind you of something else? Well, indeed this equation is very much like that for the force of gravity. A force is present only if there are two charged objects—just like in the gravitational force you need two masses. The force is inversely proportional to the distance between the two objects, just like gravity. There's also a constant there. In the case of the gravitational consonants, it's that big G we talked about a few lectures ago, the same sort of thing is true for the electrostatic force.

The law for electrostatic force, though, is different from gravity in two important ways. First, the electrostatic force can be either attractive (if you have opposite charges) or it can be repulsive (if the charges are the same), and that's different from gravity. Remember, gravity is always attractive. The second difference is that the electrostatic force is vastly greater than gravity. A comb with just a few billion extra electrons can pick up a piece of paper against the entire gravitational pull of the Earth. So your own experience with static cling with frizzy hair is that this electrostatic force is vastly greater than gravity. In these equations for gravitational energy for electrostatic forces, we use the concept of mass, and we use the concept of charge. We can measure these quantities, but their physical origins still remain obscure.

What is it about an object that gives it this ability—either the charged object or an object with mass—to exert forces on each other? That's still a very subtle point. You might think that static electricity isn't a central topic of physical research these days, and that's probably true to a large extent, but we do have to deal with the consequences of static electricity all the time. We have lightning—we have to deal with that. It occurs whenever you have agitated rain, very turbulent rain, up in the high levels of the atmosphere, and you're actually stripping electrons away from the upper levels of a cloud, and then the electrons get concentrated as the base in water droplets, and as the rain falls you're transferring electric charge from the cloud to the ground. Lightning is merely a way of compensating for that charge that's being transferred from the upper levels of the atmosphere down to the ground.

You can have cloud-to-cloud lightning in these situations. You can have cloud-to-ground lightning, and even in rare situations, ground-to-cloud lightning. So all different situations where you have an unbalance of charge leads to lightning. Often the ground itself is not going to have a net charge. You say, "Well, why would lightning go from a charged place to something with zero charge?" What happens is you have the attraction, remember, of oppositely charged particles. So, for example, if the top of the cloud is negatively charged, the ground is induced to have a positive charge. The negative charges in the tops of trees, or for example, are pushed away, and that leaves a residue of positive charge. So you induce a charge in the ground even though the ground does not have a net charge, and that's why lightning tends to hit the highest objects—church steeples, or trees, other tall objects. If you're in an open field, you better get away if lightning strikes, because you're going to be the tallest object around and lightning will look for you.

There's an important technological use of static electricity in our everyday lives, that's the Xerox machine, and let me just tell you briefly about that fascinating object, the Xerox machine. You have a rotating drum, and that drum is made of special material that takes a static charge. When light shines on part of the drum, that static charge is eliminated, that part loses its charge completely. Therefore, if you shine a black-and-white pattern onto the drum, that black-and-white pattern is reproduced in the static charges on the drum. The drum rotates by a pile of very tiny plastic beads, which are, by static electricity, attracted to the charged parts of the drums, but not to the uncharged parts of the drum. And that black powder, that black plastic, is then impressed onto a piece of hot paper, the plastic melts onto the paper and that's your Xeroxed copy. So Xerox machines use static electricity in a very basic way. By the way, if you want a darker copy, you just merely turn up the voltage. You get a higher static charge. If you want a lighter copy you turn down the voltage and you get less of that black powder attaching to the drum.

Okay, in summary, let's review what we've learned about these two seemingly unrelated phenomena—magnetism and electricity. Magnets always display two poles, designated north and south. This situation is called a magnetic dipole. When two magnets approach each other, opposite poles attract, like poles repel each other. The Earth itself is a giant dipole magnet, so magnetized compass needles point to the North Pole.

Static electricity is a force that occurs between two electrically charged objects. Negatively charged objects have an excess of electrical fluid, or electric charge, what we now called electrons. Positively charged objects have a deficiency of electrons. Opposite charges attract, like charges repel each other, and this situation leads to many phenomena—lightning, Xerox machines, and so forth. Static electricity is interesting indeed, but the true practical importance of electricity was not realized until scientists began to study electric charges in motion—the phenomenon of electricity, which is the subject of the next lecture.

Lecture Twelve
Electricity

Principle: *"Electric currents are produced by moving electrons in a closed path."*

Scope:

Most modern uses of electricity rely on electrons that move. Italian physicist Alessandro Volta (1745–1827), building on investigations related to animal electricity by the Italian anatomist Luigi Galvani (1737–1798), devised a method for applying a continuous motive force to electrons with his invention of the battery in 1794. Volta found that an electrical potential is produced if rods of two different metals are placed into an acidic chemical bath. Volta's battery marked a turning point in electrical science, because, for the first time, researchers could rely on a steady source of electricity, rather than transient sparks and discharges. The battery introduced a new era of research on electric circuits, which incorporate three components: a source of electrical energy, a device that responds to this electrical potential, and a closed loop of conducting material.

Outline

I. Static electricity is limited in its practical applications. Most modern uses of electricity rely on electrons that move. Newton's first law of motion demands that electrons can't move unless a force is applied. A method for applying such a force was discovered in 1794, when the Italian physicist Alessandro Volta (1745–1827) invented the battery.

  A. Alessandro Giuseppe Antonio Anastasio Volta was the fifth of five sons born to an impoverished family of lesser nobility in northern Italy. He showed a flair for foreign languages and began experimental studies on electricity while still a teenager. He spent his career as professor of physics at the University of Pavia.

  B. Volta's most important contributions followed discoveries of his countryman, the anatomist Luigi Galvani (1737–1798), who focused much of his work on the subject of animal electricity.

1. Galvani studied the effects of electric sparks, which caused the muscular legs of dead frogs to twitch and convulse.
2. While performing these experiments he noticed the leg would twitch when touched simultaneously by a brass wire and a steel scalpel.
3. Further experiments clarified this phenomenon. Galvani showed that when copper and iron wires were inserted into different parts of a dead frog's legs, and these two wires were then touched together, the legs would convulse.
4. The anatomist Galvani interpreted this phenomenon in terms of "animal electricity," an electricity intrinsic to biological tissues.

C. The physicist Volta directed his attention to these phenomena, focusing more on the metallic elements of Galvani's experiment than the biological components. He soon became convinced that the juxtaposition of two different metals led to the observed electrical phenomena, which he called "metallic electricity."
1. He found that different pairs of metal produced different degrees of effect.
2. A feud developed between supporters of Galvani's animal electricity and Volta's metallic electricity.

D. Volta's view soon prevailed because his electrical effects could be produced independently of frog's legs or other biological material.
1. Electrical potential is analogous to the gravitational potential of water behind a dam.
2. For Volta, the next step was to devise various arrangements by which different metals were placed in contact. He found that he could produce electricity just by stacking alternating plates of metals, such as silver and zinc, in a saltwater bath.
3. Volta's battery marked a turning point in electrical science. For the first time, researchers could rely on a steady source of electricity, rather than transient sparks and discharges.

E. Volta's battery and its successors proved invaluable in chemical experiments.

1. Within weeks of its announcement, the British chemists William Nicholson and Anthony Carlisle built a crude battery, and they used it to decompose water into hydrogen and oxygen for the first time.
2. Batteries were used to decompose other substances, leading to the discovery of several new elements.
3. Larger batteries were built with more pairs of metal plates. The flamboyant English chemist and science lecturer Humphry Davy constructed a mammoth battery with 2000 double plates at the Royal Institution. In 1810, he became the first person to demonstrate electric lighting when he vaporized charcoal, platinum, and other materials in blinding incandescent displays.

II. The battery introduced a new field of research on electric currents and electric circuits.
   A. An electric circuit incorporates three components: a source of electrical energy, a device that responds to this electrical potential, and a closed loop of conducting material.
      1. The source of electrical energy might be a battery, a solar cell, or a hydroelectric plant.
      2. An electrical device is an object or substance that responds in some interesting or useful way to the voltage of the source.
      3. A piece of wire or other conducting material is required to close the loop of an electric circuit.
   B. Electrical circuits can be quantified in several useful and important ways.
      1. The flow of electrons through a circuit is called an electric current, measured in amperes.
      2. The electrical potential that causes electrons to move is measured in volts, in honor of Volta.
      3. Every circuit has some resistance to the flow of electrons, measured in ohms.
      4. Power is defined as work divided by time. In an electrical circuit, power is the current times the voltage and is measured in watts.
   C. In homes, the energy source is typically a power plant many miles away. The separate circuits in your home that feed off this power network are of two types.

1. Series circuits have several devices, each linked up after the next in a single large loop. The same current flows through every device.
2. Parallel circuits are arranged so that a single source supplies voltage to separate loops, each dedicated to one device. Every device sees the same voltage, but different currents.
3. Kirchhoff's Laws systematize the behavior of circuits. The first law is a restatement of the law of energy conservation: energy produced by the source equals the energy consumed in the circuit (including heat energy of resistance).
4. The second law is a statement of conservation of current: the current flowing into any junction equals the sum of the currents flowing out. Since current is simply the number of electrons flowing past a point, this law is equivalent to saying that electrons are conserved.

D. The word electricity has been introduced in many different contexts: animal electricity, metallic electricity, static electricity, lightning, and so forth. British physicist Michael Faraday (1791–1867) showed that all forms produce the same effects, and thus unified the many faces of electricity.

E. The unification of electricity with the seemingly unrelated force of magnetism transformed the world's technology.

Essential Reading:
Trefil and Hazen, *The Sciences: An Integrated Approach*, Chapter 5.

Supplemental Reading:
Amdahl, *There Are No Electrons*.
Dibner, *Alessandro Volta and the Electric Battery*.
Harre, *Great Scientific Experiments*, Chapter 18.
Pera, *The Ambiguous Frog: The Galvani-Volta Controversy on Animal Electricity*.

Questions to Consider:
1. To what extent did the differing conclusions of Galvani and Volta regarding the twitching of frog's legs reflect their different scientific specialties?

2. Electricity is often described in terms of an analogy to a plumbing system. What are electrical analogues to pumps, pipes, water pressure, flow rate, and an obstruction in a pipe?

Lecture Twelve—Transcript
Electricity

We take so much for granted in this modern world. You turn on the faucet and water's going to come out. You go to the grocery store, and there are going to be stacks of oranges and grapefruit anytime of the year. You turn on the light switch and the lights are going to go on. A century ago, you wouldn't have been quite so blasé about these everyday occurrences. Indeed, all these simple modern marvels and so many more are a consequence of a great scientific principle that electric currents are produced by moving electrons in a closed path.

This lecture is divided into two parts. First, I'm going to tell you about the surprising history of the invention of the battery, which was invented in Italy more than 200 years ago. Then we're going to look at the design of electrical circuits and their principle components—that is, a source of energy, a device that does something interesting or useful, and finally a closed loop of conducting material, usually wired, and completes that circuit.

In the last lecture I introduced the idea of static electricity. Objects obtain static electricity when they have either an excess or a deficiency of electrons. So electrically charged objects then can exert forces on each other. We saw that opposite charges attract each other, like charges—two negative charged objects or two positive charged objects—repel each other. You can do lots of simple experiments to study this phenomenon and document the details of static electricity, and indeed, even come up with a quantitative description of it as in Coulomb's law, that the force of static electricity is equal to the charge of the first object times the charge of the second object divided by the distance between them squared. But static electricity, while it's intriguing, is rather limited in its practical applications. Most modern uses of electricity rely on electric charges that are electrons that are in motion.

Newton's first law tells us that if electrons are going to move, if we're going to put them into motion, you need to apply a force. It was not until 1794 when Italian physicist Alessandro Volta invented the battery that such a force was readily available for people in the laboratory, and then eventually, people in homes and businesses as well.

So at this point I first want to introduce two astonishing characters in the history of science: Volta and his fellow countryman, Luigi Galvani. Their careers were linked in this curious sort of way, and they were linked because of the behavior of frogs, of all things.

Alessandro Giuseppe Antonio Anastasio Volta, he was the fifth of five sons born to an impoverished family of lesser nobility in northern Italy. I guess what Volta lacked in material wealth, he more than made up for in middle names. At first the young Alessandro was a dull-witted child it seemed. It was not until his fourth year that he began to speak. But soon afterwards he advanced extremely rapidly in school. He showed a flair for foreign languages, of all things, even though he didn't speak when he was very young.

He began experimental studies on electricity when he was still a teenager. He gained great fame for his invention of the electrophorus; that was the device that held electrostatic charge and then was able to transfer it from one object to another. This illustrated many of the basic principles, so it was very important in its day. Shortly thereafter, he was offered a professorship in physics at the University of Pavia, and that's where he spent the rest of his career. Now Volta's most important contributions followed discoveries of his fellow countryman, the anatomist Luigi Galvani, who lived from 1737 to 1798.

Galvani focused much of his work on the subject of animal electricity—that's the kind of phenomenon as exemplified by the electric eel. But animal electricity was believed to be held by virtually all living things to one extent or another. Galvani was born in 1737 in the northern Italian city of Bologna. There he was raised and educated all the way through a medical degree. He went on to become a faculty member at the University of Bologna, and president of the Bologna Academy of Sciences. So Bologna was certainly his home. He spent his entire career there, and he died in the exact same house where he was born, of all things.

Luigi Galvani became an authority on physiology, especially on dissection, and he was fascinated by the interaction of tissues with electricity. So he liked to subject various kinds of tissues to electric charges, electric shocks and see what happens. His most famous experiments involved studying the effects of electric sparks on the muscles of legs of frogs, the very muscular legs of these frogs that you see, and that allow the frogs to leap and kick so hard. He discovered the remarkable effects of static electricity, which causes their legs to twitch and convulse spontaneously, even from a dissected frog. While performing these experiments he noticed something rather new and unexpected. What he had done is he had attached a brass wire to a frog's leg, and he was preparing to induce a spark by touching the leg with a steel scalpel.

And here's a quote from one of his papers: "While one of those who were assisting me touched lightly and by chance the point of the scalpel to the internal crural nerves of the frog, suddenly all the muscles of its limbs seemed to be so contracted that they seemed to have fallen into tonic convulsions." But the thing is, these convulsions were not caused by a spark. It was an entirely new phenomenon, because this was occurring just because it was a brass hook in the frog and a steel scalpel touching at the same time.

In another set of experiments, Galvani wanted to study the effects of lightning on severed limbs of various kinds of animals, so he suspended an iron wire across the courtyard of his villa, and then he had brass hooks, which held the limbs to the wire, while the nerves of the various organisms were attached to the iron wires directly. So here you had brass and iron again, two different metals stuck into these severed limbs, which were hanging outside. When he placed the wire outside during an electrical storm, the various limbs would jump and twitch as he expected. But curiously, the twitching sometimes occurred on dry days, on sunny days, just as long as the brass and the iron were both attached to the limbs of these frogs.

Lots more experiments were required to kind of clarify and define this phenomenon. Galvani showed that when copper and iron wires were inserted into different parts of these dead frogs' legs, the two wires were then touched together, and that's what caused the legs to convulse. You didn't need a spark. You didn't need any sort of other electrical current. You just needed two pieces of metal. This was a new kind of electrical phenomenon. The anatomist Galvani interpreted this phenomenon in terms of "animal electricity." He said this is an electricity that's intrinsic to living organisms, to biological tissues. Others wondered whether Galvani had actually discovered the life force, which had been much debated at the time.

Others had to study this phenomenon as well. Upon hearing of Galvani's experiments, the physicist Volta directed his attention to these same phenomena. So he tried the same kinds of experiments, but he focused more on the metallic elements of Galvani's experiment than the biological components. After all he was a physicist. He was interested in the physical world, and so metals seemed to be the more basic thing to study to him. He soon became convinced that it was the juxtaposition of these two different metals that led to the electrical phenomenon, which he then called "metallic electricity."

He found that different pairs of metals produced different degrees of the effect, and there's a very clear sequence. The sequence, which is known as the electrochemical series, starts with zinc at one extreme followed by tin and then lead and iron, copper, platinum, gold, and silver. So the farther apart in that electrochemical series you have two metals, the greater the effect, the greater the twitching of a frog's leg, for example.

At this point, a very unfortunate feud developed between supporters of Galvani on the one hand—the idea of animal electricity—and Volta's metallic electricity. And the lines were divided very cleanly. An extension of this conflict was between biology and physics. Which would take the primary role? And as so often happens in science, it's the physicist who seems to win out over the biologist.

Volta's view soon prevailed, because his electrical effects could be produced independently of frog's legs or any other biological material for that matter. He found that if you put two rods of different metals into an acidic bath, just a chemical bath, an electric potential would form between those two rods, so you could actually create electric sparks or other electrical phenomenon just by putting two pieces of metal into that acid bath. This electrical potential that was created is quite analogous to the gravitational potential of water behind a dam. In other words, you're storing electrical potential energy by having these two pieces of metal in an acid bath. Just as if you release some of the water, you release gravitational potential energy, so by touching the pieces of metal together you'd release some of this electrical potential energy. You can do work with those moving electrons, with those moving electric charges.

And this is the beginning of the study of electricity, the study of moving electrons. For Volta, the next step was a very logical one. It was to devise various arrangements by which you placed different metals in contact and studied various configurations. He found that he could produce electricity just by stacking alternate layers, alternate plates or discs of metals, such as silver, zinc, silver, zinc, silver, zinc, in a saltwater bath. This device was described in 1800 in Volta's article on the electricity excited by the mere contact of conducting substances of different kinds—this was the invention of the battery.

Of this battery he wrote, "The apparatus to which I allude, and which will no doubt astonish you, is only the assemblage of a number of good conductors of different kinds, arranged in a certain manner." And that certain manner was to alternate plates or discs of material, two different metals and keep them in some kind of solution, an acid bath or saltwater, or some such thing, which would allow the current to flow.

Volta's battery marked a turning point in the study of electrical sciences. For the first time, researchers could rely on a steady source of electricity, a reliable source of electricity that continuously flowed, as opposed to transient sparks and discharges—the kind produced by a friction machine, for example. Surprisingly, Volta's very first experiments were not on an inorganic physical system, but rather on the human body itself. For example, he tried his electrical effects on the tongue. He said that when you applied the two electrodes to your tongue, you sometimes tasted a strange, bitter taste. If you put it near your eye, you saw flashes of light. Pain and convulsions could also be produced if you put the electrodes in certain parts of your body. So, he experimented on humans. He also experimented on many animals. And he also wrote, for example, "it is very amusing to excite, in this manner, the chirp of a cricket."

So it's not exactly our modern thinking of how we should use electricity, but it was the type of experiment that Volta could perform. Other investigators carried these kinds of experiments to extremes. Human corpses were made to kick on a table, even to sit up when you attached batteries to the appropriate muscles of a dead human being. These bizarre demonstrations, I think, must have contributed to Mary Shelley's *Frankenstein*, which was written about the same time, in the early 1800s, and that, of course, raises real serious questions about what are the appropriate limits to scientific research. If you read Frankenstein, you realize that's one of the underlying questions of that book, and one that faces us still today.

Volta's experiments on electricity and applied to human cadavers, and so forth—that must've raised these doubts in many people's minds. Volta's experiments also contributed a great deal to the development of electromedical devices, especially home remedies, which were, by the end of the 19th century, available in huge numbers just in the Sears Roebuck catalogue, for example. You could go and you could buy home energy devices, little portable devices, where by turning a crank, you would be able to produce an electric charge, and then you'd have two electrodes, which you could place on various parts of your body, perhaps to soothe headaches or other muscle aches of various sorts. I don't know if they worked, but they certainly were sold in large numbers and it became conventional wisdom that electricity was one of the ways of curing the human body.

Volta's battery and its successor proved invaluable in chemical research as well. Within weeks of its announcement there were two British chemists, William Nicholson and Anthony Carlisle, who constructed their own very crude battery. What they did is they took 36 silver coins, silver half-crowns, and they interspersed them with 36 discs of zinc. So they had a silver-zinc battery, and they immersed this into an acid bath, and they were able, with their very crude battery, for the first time, to break water, that's H_2O down into hydrogen plus oxygen. This was a very important way for chemists to understand the constituents of nature. Batteries were also used to decompose many minerals and other substances. This led to the discovery of many new elements, as we'll see in Lecture Twenty.

Larger and larger batteries were built with more and more pairs of metal plates, larger and larger metal plates. The flamboyant English chemist and science lecturer Humphrey Davy constructed a mammoth battery with 2,000 double plates. He did this at the Royal Institution, to make a huge public demonstration. This array was extremely dangerous; it filled an entire great hall, and of course produced a huge voltage, a very, very large amount of electrical energy. In 1810, Davy became the first person to demonstrate electric lighting when he vaporized rods of charcoal and platinum and other material in blinding incandescent displays. So he'd hook up his battery to these pieces of material high above the auditorium, or high above the floor and they'd burst into flame and create a blazing light for as long as they were able to before they just vaporized.

When you pass a strong current through any material, it can heat up, and the same principle is used today in electric arc welding, as well as, of course, in light bulbs, where you pass the current through a very thin filament, which then glows very, very hot.

Let's talk about electric circuits, because that's where we really use electricity in our lives today. The battery introduced a new field of research on electric currents and these things called electric circuits. An electric circuit is a very simple concept. It incorporates three different components. Every electric circuit is like this. First of all, you need to have a source of electrical energy. You need to have a power plant or a battery or something else that generates electricity—allows electrons to move. So you have to have that electric force to push electrons along. The second thing you always need is a device; a device is something that does something interesting, does something useful when you put electric current through it. A light bulb, a heating element, a semiconductor array that operates your radio or your television, or some other electronic device—those are the devices. And finally, you need to have a closed loop of conducting material. Usually, the closed loop is completed with a piece of wire.

So let's look at these different components more carefully. That source of electrical energy—of course, you can get a battery, you can get a solar cell, a hydroelectric plant. You always have to have a place where there's a positive terminal and a negative terminal, and charge then can flow from one to the other. This push of electric charge is called voltage, and you measure this potential in volts.

The second component is the device. This word *device* has a very special meaning in the study of electricity. A device is something that uses that electric energy to do something. It responds in some interesting or useful way to that electric current. In the case of 19th-century physicists, most devices were designed to display some interesting phenomenon, for example, in a classroom environment, or in a laboratory environment. So decomposing water, essentially water becomes the device in that case, something that glows, even a frog's leg can be thought of as the device in those early experiments. But today, of course, devices are anything that you'd plug into a wall socket and use with electricity.

Then you have to have a piece of conducting material, and while it's usually a wire, there are certainly other kinds of materials that can close a loop, and if you look inside your television, you'll see various strips of metal that have been deposited onto a plastic surface—they may be the conducting material, or even in some cases, the chassis of a device becomes part of the closed circuit, and we'll see how that works. A very simple everyday example of a device of this sort is a flashlight, which incorporates all three of these components. If you could unscrew your flashlight, you see the source. In this case, it's two batteries. Typical batteries of this sort are about 1.5 volt, so two batteries together provide you 3 volts of current. The device is the light bulb, which occurs at the end of the flashlight. Current flows into this light bulb. It goes through a very tiny filament, and that filament, because of electrical resistance, heats to a very high temperature and, therefore, glows brightly. And finally, the circuit is completed by a strip of metal that goes down the side, the barrel of the flashlight. You also see a coil of wire, usually at one end of the flashlight, and at the other end you'll see the contact points for the battery and the other strip of wire, and so you have a complete circuit when you put the whole flashlight together.

Flashlights and most other electrical appliances also have a switch. A switch is merely a device that, when you open and close it, you're breaking the continuous loop of conducting material. When the switch is open you don't have current flowing, but when the switch is closed you do. Basically, all circuits are of this sort. Even the circuit you plug into the wall, there's a continuous loop of wire that extends from your home all the way to the power plant. That's the complete circuit. The flow of electrons through an electrical circuit is called the electrical current. Current is measured in amperes after the French physicist, Andre Marie Ampere, who lived from 1775 to 1836. One ampere corresponds to about 6 billion billion electrons passing a point in that circuit every second. So, you're literally talking about the flow of electrons. Current is the flow, it's the movement of electrons and as 6 billion billion electrons pass a point in the wire every second, you have a current of one ampere.

You've probably used this term in conjunction with fuses or with circuit breakers. We use fuses in our homes because if you have too much current being pulled through a wire, the wires can get hot. So in order to protect your house from burning down, you put a fuse, you put a circuit breaker in the way. A fuse is designed to burn up if the current gets too high. So, for example, in a 30 amp fuse, if the current gets up to be 35 or 40 amps for any extended period of time, the fuse will burn out and then you'll have to replace that fuse, but that's better than having to replace your entire house, I suppose.

Then we come to the electrical potential. You have to be able to push electrons through your wires, and you do that with the electric potential, which is measured in volts, in honor of Volta. Remember, voltage is a form of potential energy. It's the energy, this ability to exert a force over distance, and you're pushing a force on those charged particles, on those electrons. The higher the voltage, the bigger the acceleration of your electrons. Typical flashlight batteries are 1.5 volts. You can by hobby batteries that are 9 volts or 6 volts sometimes. The typical voltage in a wall socket could be 115 volts, for example. So we have lots of different voltages and you will often see in your appliances that they are rated for certain voltages.

A third component of every circuit is resistance. Every circuit has some resistance to the flow of electrons. Electrons collide with other electrons and the atoms that make up the wire, and they thus convert some of their energy to heat. This is part of the second law. You just can't have any transfer of energy from one form to another without inevitably losing some of that energy as heat, and this is a case in point. As electrons flow through the wire, some of that energy that you're paying for in your electric bills is converted to the heat, just in the wires. Resistance is measured in ohms, that's after the German physicist George Ohm, who lived from 1787 to 1854. Once again this classic period of the study of electricity, the first half of the 19th century, and that's when Ohm did most of his important work. Ohm discovered a very simple relationship among current voltage and resistance. In fact, that is, current is proportional to voltage and inversely proposal to resistance, or $V = I \times R$. Voltage equals current times resistance. You can actually buy a device in most hardware stores called a multimeter. A multimeter can measure resistance; it can measure a current; it can measure voltage of any particular set of components in your electrical device. It's an extremely useful and compact device.

We need one other term associated with electricity, and that's power. Recall that power is defined as work divided by time. In an electrical circuit then, power is the current times the voltage. It's measured in watts. If you see a light bulb, if you see amplifiers, they're always rated in watts. That's the amount of power that's consumed. The higher the wattage, the faster the energy is consumed by that object, be it a light bulb or an amplifier or whatever. Power equals current times voltage. In our homes, the energy source is typically a power plant that's many, many miles away. A few homes have their own portable gasoline generators in case there's a power failure, but most of our electric energy comes from remote sources. Electricity is carried over power lines and these power lines link most of the United States in one gigantic network, one gigantic circuit that's all interconnected. This physical linkage, I think, is one of the most dramatic and transforming events of the last century. The separate circuits in your home feed off this one great network that's nationwide.

Let me tell you about two kinds of different circuits that you find in your home and in other common devices. These are called series circuits and parallel circuits. In series circuits, you have several devices, each of them linked up one after another after another in just a single large loop. The same current flows through every device in that circuit, though the different devices in general are going to have different voltages across them because the voltage changes as you go along that circuit. If any one of the devices in a series circuit is broken, the whole circuit fails. And I can show you that effect in a little device here.

Here we have three light bulbs connected in a series, in just one loop of wire connected to a battery. If you unscrew one light bulb, the whole circuit fails. Screw that light bulb back in and it works. You remember Christmas tree light chains that are like this. One light burns out and the whole chain of lights is useless. That's a series type of circuit.

The other kind of circuit is called parallel circuit, and in this case the different devices are arranged so that a single source supplies voltage to separate loops of wire. Every device sees exactly the same voltage, but in general the different devices are going to see different currents. Each device is going to work even if the other ones fail, so there's a slightly different strategy here. You can show this in the same way. Here are two light bulbs linked up in parallel. Two lights go on. If I unscrew one light, the other one works. That's a parallel circuit, and that's the way modern Christmas tree lights are usually done. So even if a single light burns out, you don't have to just throw away the whole strand.

There's a way of systematizing all these relationships between circuits that are called Kirchhoff's Laws—extremely important in electrical engineering—the systematized behavior of circuits. The first law is just a restatement of the law of the conservation of energy. It says, "The energy produced by the source (a battery, for example) equals the energy consumed in the circuit, including the heat that is lost as a result of resistance."

The second law then is the statement of the conservation of current in an electrical circuit. It says, "The current flowing into any junction equals the sum of the currents flowing out of that junction." Now this is pretty logical because current is just a matter of electrons flowing through the wires, and the number of electrons flowing into a junction equals the number of electrons flowing out of that junction.

This word *electricity* has been introduced in so many contexts here. We've talked about animal electricity. We've talked about metallic electricity, static electricity, lightning, and so forth and so on. But you might ask yourself, "Are these different forms of electricity the same, or are they somehow fundamentally different?" That was another question addressed in the 19th century. And it was Michael Faraday, the great scientist Michael Faraday, who we've talked about and we'll see more in the next lecture, who made very careful systematic surveys of all these different kinds of electricity. And what he did, he was able to show that all of these different forms of electricity produce exactly the same kind of phenomenon. They all produce sparks, if you treat them in the right way. They all can flow through wires. They all can be made to do work. And indeed, it was Faraday's research that for the first time showed that animal electricity of an electric eel, the electricity coming from a battery, the electricity of lightning—all these phenomena were one and the same, and indeed we now know they all result from the movement of electrons.

Let me summarize this lecture on electricity. The science of technology of this phenomenon of electricity was begun by the development of the battery more than 200 years ago. It was initially nothing more than a stack of alternating plates of metal, but the battery provided a steady source of electrical energy that could be used to move electrons from one place to another. It could be used to make an electrical circuit. And we saw that there are three components of all electrical circuits. First you have to have a source of energy, like a battery or a power plant. And then you have to have a device like a light bulb or a heating element. And finally, you need to have a complete loop of conducting material, usually made of wire. These were great discoveries, but it was the great and startling unification of electricity with magnetism that transformed the world of technology, and that's the subject of the next lecture.

Concordance to the Science Content Standards of the *National Science Education Standards*

The *National Science Education Standards* (National Academy of Sciences, 1997) represents a consensus among thousands of scientists and educators, regarding the most effective approaches for teaching and learning about science. The *Standards* provide American parents, teachers, and school administrators with an unprecedented building code for developing effective science curricula. The *Standards* focus on several aspects of science education, including classroom methods, assessment, teacher preparation, and science content standards.

This 60-lecture course has been designed to introduce and review all of the scientific principles that are included in the K–12 "Content Standards" portions of the *National Science Education Standards*. All content standards in physical, earth, space, and life sciences are covered, as are aspects of the nature of science, the history of science, science and technology, and science in personal and social perspectives.

For the convenience of curriculum developers, the following outline matches all *National Science Education Standards* content standards to specific lectures (numbered 1 to 60).

I. Unifying Concepts and Processes, K–12
   A. Systems, order, and organization (all lectures, especially 1–4)
   B. Evidence, models, and explanation (all lectures, especially 1–4)
   C. Constancy, change, and measurement (1, 3, 35–42, 54–58)
   D. Evolution and equilibrium (35–36, 44, 54–57)
   E. Form and function (22–26, 43–47)

II. Content Standards, K–4
   A. Science as Inquiry (1–2, 60)
   B. Physical Science:
      1. Properties of objects and materials (22–26)
      2. Position and motion of objects (3–6)
      3. Light, heat, electricity, and magnetism (11–16)

- **C.** Life Science:
    1. Characteristics of organisms (43–44)
    2. Life cycles of organisms (44, 60)
    3. Organisms and environments (58–59)
- **D.** Earth and Space Science:
    1. Properties of earth materials (39–42)
    2. Objects in the sky (3–4, 29–31)
    3. Changes in earth and sky (3–4, 29–32)
- **E.** Science and Technology:
    1. Abilities of technological design (many lectures, especially 52–53)
    2. Understanding about science and technology (examples throughout)
- **F.** Science in Personal and Social Perspectives:
    1. Personal health (15, 45–46, 58–59)
    2. Characteristics and changes in populations (58)
    3. Types of resources (42)
    4. Changes in environments (40–41, 58–59)
    5. Science and technology in local challenges (throughout, especially 59)
- **G.** History and Nature of Science: Science as a human endeavor (throughout)

III. Content Standards: 5–8
- **A.** Science as Inquiry (throughout, especially 1–2, 60)
- **B.** Physical Science:
    1. Properties and changes of properties in matter (22–28)
    2. Motions and forces (4–6, 11)
    3. Transfer of energy (7–8, 21, 28)
- **C.** Life Science:
    1. Structure and function in living systems (43–47)
    2. Reproduction and heredity (48–51)
    3. Regulation and behavior (44, 60)
    4. Populations and ecosystems (58–59)
    5. Diversity and adaptations of organisms (44, 57–59)
- **D.** Earth and Space Science:
    1. Structure of the earth system (36–39)
    2. Earth's history (35–39, 42)
    3. Earth in the solar system (34–36)

- **E.** Science and Technology (throughout, especially 26, 28)
- **F.** Science in Personal and Social Perspectives:
    1. Personal health (15, 27)
    2. Populations, resources, and environments (40–42, 58–59)
    3. Natural hazards (37–39, 41)
    4. Risks and benefits (15, 28, 58–59)
    5. Science and technology in society (22, 26, 28, 52–53, 57–59)
- **G.** History and Nature of Science:
    1. Science as a human endeavor (1–3, 60, and throughout)
    2. Nature of science (1–2, 33, 60, and throughout)
    3. History of science (examples used throughout)

IV. Content Standards: 9–12
- **A.** Science as Inquiry (1–2, 60, and throughout)
- **B.** Physical Science:
    1. Atoms (17–20, 27)
    2. Structure and properties of matter (17–18, 21–26, 33)
    3. Chemical reactions (21, 24)
    4. Motions and forces (3–6)
    5. Conservation of energy and increases in disorder (8–10)
    6. Interactions of energy and matter (14–16, 24, 28, 33)
- **C.** Life Science:
    1. Cells (47, 49)
    2. Molecular basis of heredity (49–52)
    3. Biological evolution (54–57)
    4. Interdependence of organisms (43–44, 58–59)
    5. Matter, energy, and organization in living systems (10, 12, 43–47, 58)
    6. Behavior of organisms (43–44)
- **D.** Earth and Space Science:
    1. Energy in earth systems (28, 30, 36, 38–39)
    2. Geochemical cycles (40–42)
    3. Origin and evolution of the earth system (34–36, 42)
    4. Origin and evolution of the universe (31–32)
- **E.** Science and Technology:
    1. Technological design (26, 28, 52)

2. Understanding about science and technology (28, examples throughout)
F. Science in Personal and Social Perspectives:
  1. Personal and community health (27, 45–46, 53, 59)
  2. Population growth (40, 58–59)
  3. Natural resources (40–42)
  4. Environmental quality (40–41, 58–59)
  5. Natural and human-induced hazards (28, 39, 41)
  6. Science and technology in local, national, and global challenges (27–28, 58–59)
G. History and Nature of Science:
  1. Science as a human endeavor (1–4, 33, 60)
  2. Nature of scientific knowledge (1–2, 60)
  3. Historical perspectives (examples used throughout)

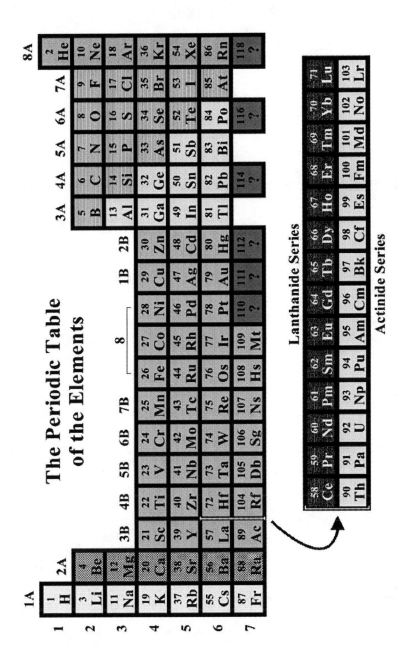

Timeline

c.3000 B.C.	Builders of ancient monuments, such as Stonehenge in England, recognize reproducible events in the heavens.
c.430 B.C.	Democritus of Abdera advocates the atomic theory of matter on philosophical grounds.
c.370–330 B.C.	Aristotle's teaching and writings on astronomy, physics, and biology exert a great influence on subsequent scholars.
c.50–59	Pliny the Elder catalogs thousands of "facts" in his 37-volume *Natural History*.
c.145	Ptolemy of Alexandria proposes an Earth-centered model of the solar system that incorporates epicycles.
1543	Copernicus publishes Sun-centered model of the solar system, and Vesalius publishes his study of human anatomy.
1572	Tycho Brahe discovers a new star—a supernova in the constellation Cassiopeia.
1581	British instrument maker Robert Norman publishes *The Newe Attractive*, in which he describes magnetic dip.
1600	English physician William Gilbert publishes *De Magnete*.
1600	Johannes Kepler becomes Tycho Brahe's assistant.
1609	Galileo Galilei builds his first telescopes. His first observations

	were published in *The Starry Messenger* in 1610.
1619	Kepler publishes *Harmony of the World*, which introduces his third law of planetary motion.
1632	Galileo publishes *Dialogue Concerning Two World Systems*, which led to his heresy trial in the following year.
1660	The Royal Society of London, the first scientific society, is founded.
1665	British scientist Robert Hooke publishes *Micrographia*, in which he describes the cells of plants.
1665–1666	Isaac Newton develops calculus, deduces the laws of motion, derives a mathematical description of gravity, and conducts key experiments in optics.
1677	Anton van Leeuwenhoek uses the microscope to discover single-celled organisms.
1687	Newton's great work *Principia* is published.
1743	Benjamin Franklin helps establish the American Philosophical Society—America's first scientific society.
1747–1752	Franklin's electrical experiments lead to the invention of the lightning rod and the theory of electricity as a single fluid.
1785	Scottish geologist James Hutton publishes *Theory of the Earth*, in which he proposes that geological

changes occur gradually over immense spans of time.

1796 .... French mathematician Pierre Simon Laplace proposes the nebular hypothesis of the origin of the solar system.

1797 .... Benjamin Thompson illustrates the mechanical equivalence of heat.

1799 .... Alessandro Volta invents the electric battery.

1808–1827 .... English meteorologist John Dalton presents the atomic theory in *A New System of Chemical Philosophy*.

1820 .... Danish physics professor Hans Christian Oersted discovers that electricity can produce magnetic fields.

1828 .... German chemist Friedrich Wohler synthesizes urea.

1831–1836 .... British naturalist Charles Darwin serves as the naturalist aboard the *HMS Beagle*.

1859 .... Darwin publishes *On the Origin of Species*.

1869 .... Russian chemist Dmitri Mendeleev proposes his periodic table of the elements.

1873 .... James Clerk Maxwell systematizes electromagnetic phenomena and predicts electromagnetic radiation.

1895 .... German physicist Wilhelm Konrad Roentgen discovers x-rays.

1898 .... Polish chemist Marie Curie, assisted by her husband Pierre, isolates the first of the radioactive elements.

1899–1904	British physicist Ernest Rutherford makes fundamental discoveries regarding the nature of radioactivity.
1902	American biologist Walter Sutton discovers egg and sperm have matching pairs of chromosomes.
1905	Albert Einstein publishes three fundamental discoveries: the atomic origin of Brownian motion, the quantum nature of radiative energy, and the theory of special relativity.
1911	Dutch physicist Heike Kamerlingh-Onnes discovers superconductivity in a sample of mercury at 4 degrees kelvin.
1924	American astronomer Edwin Hubble discovers that galaxies are immense collections of billions of stars.
1932	Particle physicist Carl Anderson discovers antimatter.
1945	American scientists develop the atomic bomb.
1952	American biochemist James Watson and British crystallographer Francis Crick solve the double helix structure of DNA.
1953	Chemists Stanley Miller and Harold Urey synthesize amino acids in a chemical experiment designed to mimic the early Earth's atmosphere and ocean.
1959	Cornell physicists Giuseppe Cocconi and Philip Morrison propose a search for extraterrestrial intelligence.

1964	Molecular biologists crack the genetic code of DNA and RNA.
1965	American oceanographer Drummond Matthews and British geophysicist Frederick Vine report on strips of magnetically-aligned rock on either side of mid-ocean ridges.
1969	American astronauts land on the Moon and return samples from the lunar surface.
1972	American paleontologists Stephen Jay Gould and Niles Eldridge propose the theory of evolution by punctuated equilibrium.
1977	American oceanographer Jack Corliss discovers ecosystems at volcanic vents a mile or more deep on the ocean floor while diving in the submersible Alvin.
1980s	NASA's Voyager 1 and 2 spacecraft visit the outer planets of the solar system.
1985	A region over the Antarctic with seasonally reduced ozone, now called the ozone hole, is discovered by British scientists.
1986	The first of a new class of high-temperature superconductors with ionic bonds is discovered by IBM scientists.
1990	NASA's Hubble Space Telescope is launched.
1996	British biologist Ian Wilmut clones a sheep named Dolly from an adult cell.

Glossary

acid rain: the formation of acidic rain drops by reaction with chemicals produced by burning fossil fuels.

alleles: different forms of a gene, some of which may be dominant and some, recessive.

alternating current (AC): electricity produced by a generator, in which electrons move back and forth.

amino acids: molecular building blocks of proteins.

animals: multicellular organisms that obtain their energy and raw materials from the biomolecules of other organisms.

antimatter: particles that, when combined with their oppositely charged matter particles, annihilate to form energy.

astronomy: the science of collecting, analyzing, and interpreting photons from space.

atom: a submicroscopic particle from which solids, liquids, and gases are made.

atomic number: the number of protons in an atom, which defines the element.

battery: a device that applies a continuous motive force to electrons.

big bang theory: the theory that proposes that the universe came into existence at one moment in time, and subsequently has undergone rapid expansion.

biology: the study of living systems.

black hole: the collapse of the remnants of a massive star into a point, from which even light cannot escape.

cancer: a disease that occurs when defects in the genetic machinery cause a cell to divide again and again to form a tumor.

carbohydrates: energy-rich molecules composed of carbon, hydrogen, and oxygen; the most abundant biomolecules on Earth.

cell: the basic unit of all living things.

Cenozoic Era: the period of Earth history from 65 million years ago to the present; the age of mammals.

chemical reactions: the breakdown or rearrangement of atoms into different substances.

chemistry: the study of atomic interactions.

chromosome: the structure in a cell that carries genes in the chemical DNA.

climate: a long-term average of weather for a given region.

computer: a machine that stores and processes information.

conduction: the movement of heat by atom-to-atom contact.

convection: the movement of heat by transfer of a mass of fluid.

convergent boundary: a plate boundary where two plates move together.

core: the inner metallic layers of the Earth.

covalent bond: a chemical bond in which two or more atoms share electrons.

crust: the outer layer of the solid Earth.

deoxyribonucleic acid (DNA): the chemical that carries genetic information.

direct current (DC): electricity produced by a battery, in which electrons flow in one direction.

divergent boundary: a plate boundary where two plates move apart and new crust is formed.

earth science: the study of our planet's history and present dynamic state.

earthquakes: sudden Earth movements that result from the gradual buildup of stress and subsequent fracture between two blocks of rock.

ecosystems: complex communities of organisms and their physical environment.

electric circuit: a system that incorporates a source of electrical energy, a device that responds to this electrical potential, and a closed loop of conducting material.

electricity: the motion of electrons in a closed circuit.

electromagnetic radiation: a form of wave energy produced whenever an electric charge accelerates; travels at the speed of light.

electromagnetic spectrum: the continuum of all possible wavelengths of electromagnetic radiation, including radio, microwaves, infrared radiation, light, ultraviolet, x-rays, and gamma rays.

electron: subatomic particle that carries a negative electric charge and participates in chemical bonding.

element: a substance that cannot be broken down into other substances by any ordinary physical or chemical means.

energy: the ability to do work.

entropy: the ratio of heat energy over temperature; a measure of the disorder of a physical system.

eukaryotes: single-celled organisms with a nucleus and other organelles.

evolution: the process by which life has changed over billions of years of Earth history.

extinction: the disappearance of a species.

fission reactions: nuclear reactions that split an atom.

force: the phenomenon that causes an object to accelerate.

fossil: any evidence of ancient life; usually preserved in rock.

fossil fuel: a carbon-based fuel obtained from the Earth, including coal, petroleum, and natural gas.

fungi: organisms that resemble plants in terms of their cell structure and growth patterns, but are nonphotosynthetic.

fusion reactions: nuclear reactions that combine two nuclei, usually hydrogen.

galaxies: collections of billions of gravitationally bound stars.

genetic disease: a disease that arises from a defective gene.

genetic engineering: the process of consciously altering a coded sequence of DNA or RNA to produce an organism with new characteristics.

genetics: the study of the ways by which biological information is passed down from parents to offspring.

gravity: an attractive force that exists between any two masses.

greenhouse effect: the warming of the Earth's surface by atmospheric gases, notably carbon dioxide, that trap infrared radiation.

Human Genome Project: a project that will provide a detailed map of the distribution of genes on the 23 human chromosomes, and the sequence of bases.

hydrocarbons: compounds of carbon and hydrogen.

igneous rocks: all rocks that form from a molten state.

ionic bond: a chemical bond that forms through an exchange of one or more electrons.

isotope: an atom for which the number of protons and neutrons are known.

istocacy: an earlier geological concept that held that mountains were great rafts of relatively light material buoyed up like icebergs on the ocean.

laser: a device that emits an intense narrow light beam of a single wavelength (an acronym for Light Amplification by Simulated Emission of Radiation).

leptons: a class of six particles, including electrons and neutrinos, that do not occur in the atom's nucleus.

Linnaean system: the system of nomenclature that assigns a specific name to each kind of organism.

lipids: biomolecules including fats, oils, and waxes; building blocks of cell membranes.

lithosphere: the strong rock layer that includes the crust and the top part of the mantle of the Earth; it is relatively thin, cold, and brittle and is typically between 50–100 km in thickness.

mantle: rocky layers of the Earth interior; mantle convection drives plate tectonics.

mass: an object's tendency to resist an acceleration.

mass extinction: a time in geological history when a large percentage of species become extinct.

Mesozoic Era: the period of Earth history from 248 to 65 million years ago; the age of dinosaurs.

messenger RNA: the molecule that copies the base sequence of a DNA segment (a gene) letter by letter.

metabolism: the cell's process of obtaining and using energy from its surroundings.

metallic bond: a chemical bond that forms when all atoms release electrons, creating a "sea" of negatively charged electrons with positively charged atoms interspersed.

metamorphic rocks: all rocks whose mineralogy is altered by the effects of temperature and pressure.

microchip, or **integrated circuit:** a semiconductor device that may incorporate thousands of transistor-like regions.

natural selection: the theory that life evolves by the gradual, selective transmission of desirable traits from one generation to the next.

nebular hypothesis: a widely accepted model for the formation of stars, including the solar system.

neutrons: electrically neutral nuclear particles with mass slightly greater than that of a proton.

nuclear reactor: a device that produces electrical energy by sustained nuclear fission reaction.

nucleus (of atoms)**:** tiny object that carries most of an atom's mass.

nucleus (of cell)**:** the central organelle of eukaryote cells; contains the cell's DNA.

organelles: discrete internal structures in a eukaryotic cell; the nucleus is an example.

organic chemistry: the field of chemistry devoted to carbon compounds.

ozone hole: a region over the Antarctic with seasonally reduced ozone.

ozone layer: a region of the stratosphere, containing small amounts of the gas ozone (O_3), which absorbs much of the Sun's harmful UVB ultraviolet radiation.

paleoclimatology: the study of ancient climates.

Paleozoic Era: the period of Earth history from 543 to 248 million years ago, characterized by the appearance of animals with hard parts.

phase transformation: a change in state or atomic structure, often resulting from changes in temperature or pressure.

photon: an individual packet of electromagnetic radiation.

physics: the study of matter in motion.

plants: multicellular organisms that obtain their energy from the sunlight.

plastic: a solid formed from complexly intertwined polymer strands.

plate tectonics: the theory that the surface of the Earth is divided into about a dozen thin, brittle, mobile plates.

polymer: a large molecular structure composed of chains of small molecules.

polymerase chain reaction (PCR): a technique to duplicate a specific strand of DNA.

Precambrian: the period of geological history before 543 million years ago.

prokaryote: single-celled organisms without any well defined internal structures, such as a nucleus.

proteins: chemical workhorses of life, built from chains of amino acids, their structure determines their function.

proton: positively-charged nuclear particle, present in all atoms.

pulsar: a neutron star that emits brief sharp pulses of energy as opposed to the steady release of energy typically encountered.

punctuated equilibrium: the theory of evolution that claims species change in relatively sudden bursts.

quantum mechanics: the study of motions at the scale of quantum jumps.

quarks: a class of six different particles that combine in twos or threes to form neutrons, protons, and other nuclear particles.

radiation: the movement of heat by electromagnetic radiation; also the energetic particles produced by radioactive atoms.

radioactivity: the spontaneous release of nuclear energy from an atom.

relativity: the theory that the laws of nature are the same in every reference frame.

ribonucleic acid (RNA): the molecule that transforms DNA into proteins.

scientific method: a cyclic process of inquiry based on observations, synthesis, hypothesis, and predictions that lead to more observations.

sedimentary rocks: all rocks that are deposited in layers.

seismology: the study of the Earth with sound waves.

semiconductors: materials that conduct electricity, but not very well.

SETI: the search for extraterrestrial intelligence; involves looking for characteristic radio signals from nearby Sun-like stars.

solar system: all objects that are gravitationally bound to the Sun.

spectroscopy: the study of light-matter interactions.

states of matter: solid, liquid, gas, and plasma, which are manifestations of submicroscopic organization of atoms.

superconductors: materials that conduct electricity without any resistance.

supernova: the sudden collapse and subsequent explosion of a star.

taxonomy: the formalized procedure for classifying and naming life forms.

transfer RNA: the molecule that matches three bases on messenger RNA to an amino acid.

transform boundary: a plate boundary where two plates move past each other.

virus: a strand of genetic material (DNA or RNA) surrounded by a coating of proteins; viruses can take over a cell's genetic machinery.

volcano: a mountain or other feature that forms when molten rock erupts as lava and accumulates at the surface.

waves: a way to move energy without moving mass.

weather: the state of the atmosphere at a given time and place.

work: the exertion of a force over a distance.

Biographical Notes

Aristotle (384–322 B.C.). Athenian philosopher whose writings on mathematics, physics, and biology, and development of the inductive method, were influential for more than 1,800 years.

Niels Bohr (1885–1962). Danish physicist who proposed an atomic model in which electrons adopt specific energies and shift from one energy to another in quantum jumps.

Tycho Brahe (1546–1601). Danish astronomer who advanced the field principally by designing, constructing, and using instruments that greatly increased the precision and accuracy of astronomical observations.

Henry Cavendish (1731–1810). British physicist who determined experimentally the value of the gravitational constant, and thus the mass of the Earth.

Nicolas Copernicus (1473–1543). Polish astronomer who devoted much of his life to developing a mathematical model of the solar system in which the Earth and other planets orbit the Sun.

Charles Coulomb (1736–1806). French physicist who determined the force law between two charged objects: Force equals the product of the two charges divided by the square of the distance between them, times an appropriate constant.

Francis Crick (1916–). British crystallographer who, in 1952, worked with James Watson to solve the DNA structure.

Marie Sklodowska Curie (1867–1934). Polish-born chemist who spent much of her career working with her husband Pierre in Paris. The Curies refined tons of high-grade uranium ores to extract small quantities of the previously unknown elements polonium and radium.

John Dalton (1766–1844). English meteorologist who presented the first statement of the modern atomic theory—that matter is composed of atoms of perhaps several dozen varieties that differ in their weights and sizes.

Charles Darwin (1809–1882). British naturalist who, after serving as naturalist on the voyage of the *HMS Beagle* from 1831–1836, developed the theory of biological evolution by natural selection.

Humphry Davy (1778–1829). British chemist who pioneered the use of the battery to isolate chemical elements.

Democritus of Abdera (c.460–370 B.C.). Greek philosopher who developed a philosophical rationalization for atoms, which are indestructible, but may be rearranged to form different substances.

Albert Einstein (1879–1955). German physicist who made several fundamental contributions to science. In 1905, alone, Einstein demonstrated the atomic origin of Brownian motion, provided compelling evidence for the quantum theory of matter, and produced the first installment of his theory of relativity.

Michael Faraday (1791–1867). British physicist who, in 1831, discovered electromagnetic induction and inventor of the electric generator.

Benjamin Franklin (1706–1790). Famous American statesman and signer of the Declaration of Independence, who devised an explanation of electrical phenomenon.

Rosalind Franklin (1920–1958). British crystallographer and chemist who obtained the first x-ray photographs of DNA in the early 1950s.

Galileo Galilei (1564–1642). Galileo transformed both the content and the methodology of science. He was a pioneer in the design of elegant experiments to study the physics of motion, and he was the first astronomer to use the telescope; his discoveries ultimately led to his heresy trial in 1633.

Luigi Galvani (1737–1798). Italian anatomist who noticed that the leg of a dead frog would twitch when two different metals were touched to each other and to the exposed nerves of the leg, even when no electric shock was applied.

William Gilbert (1544–1603). English physician and physicist who systematized his own magnetic investigations with earlier work to show that the Earth, itself, is a giant magnet with its own field.

Werner Heisenberg (1901–1976). German physicist who expressed the uncertainty principle, which states that you can't know the exact position and velocity of an object at the same time.

Robert Hooke (1635–1702). British physicist and microscopist who discovered units of plants, which he called "cells."

Edwin Hubble (1889–1953). American astronomer who, in 1924, discovered that galaxies are immense collections of gravitationally bound stars far outside our own Milky Way galaxy. He observed that many galaxies are receding at velocities proportional to their distance.

James Hutton (1726–1797). Scottish geologist who proposed the doctrine of uniformitarianism, that great geological changes take place through countless decades of gradual increments.

James Prescott Joule (1818–1889). British physicist who helped to develop the first law of thermodynamics and the mechanical theory of heat.

Johannes Kepler (1571–1630). Tycho Brahe's mathematically gifted assistant, who analyzed data for Mars and derived three laws of planetary motion.

Pierre Simon Laplace (1749–1827). French mathematician who developed the generally accepted model for star formation by gravitational attraction of dust and hydrogen gas into an ever denser, more compact cloud, which flattens into a rotating disk.

Antoine Lavoisier (1743–1794). Influential French chemist, who contributed to the understanding of oxidation and stated its importance in respiration. Lavoisier favored the caloric theory of heat.

Anton van Leeuwenhoek (1632–1723). Amateur Dutch scientist who was the first to make extensive use of the microscope in the 1670s, and who discovered the abundance of microscopic life.

James Clerk Maxwell (1831–1879). Scottish physicist who presented four equations that codified every aspect of electromagnetism, including the previously unrecognized phenomenon of electromagnetic radiation.

Johann Gregor Mendel (1822–1884). Czechoslovakian botanist and monk who was the founder of classical genetics. Mendel developed his laws of heredity during more than 28,000 separate cross-breeding experiments on pea plants.

Dmitri Mendeleev (1834–1907). Russian chemist who, in 1869, systematized the weights and chemical properties of 63 chemical elements in his periodic table of the elements.

Isaac Newton (1642–1726). British natural philosopher and mathematician who made fundamental discoveries in several branches of study. During the remarkable period of 1665–1666, Newton developed calculus, the laws of motion, the law of universal gravitation, and principles of optics and light. Many of his ideas were summarized in the Principia of 1687.

Robert Norman (c.1550–1600). British sailor and instrument maker who described the dip of compass needles. His work foreshadowed the concept of a magnetic field.

Hans Christian Oersted (1777–1851). Danish professor of physics who, while lecturing in front of a classroom in 1820, discovered that electricity can produce magnetic fields.

Louis Pasteur (1822–1895). French chemist who debunked the prevailing idea of spontaneous generation. Pasteur's dictum of "No life without prior life" pushed back origins to an inconceivably remote time and place.

Max Planck (1858–1947). German physicist who, in 1900, proposed the idea that energy comes in discrete bundles, called "quanta," at the atomic scale. This theory helped to explain the spectrum of electromagnetic radiation emitted by a "black body" that absorbs all electromagnetic radiation that falls upon it.

Pliny the Elder (23–79). Roman scholar who catalogued thousands of "facts" that were known to him or sources he deemed to be reliable, in his 37-volume *Natural History*.

Ptolemy of Alexandria (c.100–170). Greek astronomer who proposed an Earth-centered model that incorporated circular orbits modified by secondary circles, called epicycles.

Ernest Rutherford (1871–1937). New Zealand-born British physicist whose studies of radioactivity led to experiments that demonstrated the existence of the atomic nucleus.

Benjamin Thompson (1752–1814), known as Count Rumford. American-born inventor who demonstrated that heat is a form of mechanical work and thus is equivalent to energy.

William Thompson (1824–1907), known as Lord Kelvin. British physicist who made significant contributions to understanding the laws of thermodynamics.

Alessandro Volta (1745–1827). Italian physicist who invented the battery in 1794.

James Watson (1928–). American biochemist who, in 1952, worked with Francis Crick to solve the DNA structure.

Bibliography

This lecture series draws heavily on three books, which are suggested as companion texts. These volumes provide the essential background reading for all lectures.

Hazen, R. M. and M. F. Singer *Why Aren't Black Holes Black: The Unanswered Questions at the Frontiers of Science*. New York: Anchor, 1997.

National Research Council *The National Science Education Standards*. Washington, DC: National Academy Press, 1996.

Trefil, J. S. and R. M. Hazen *The Sciences, An Integrated Approach*, 2nd edition. New York: Wiley, 1997.

In addition to these essential readings, a number of textbooks and other volumes provide more detailed overviews of specific fields.

Amato, I. *Stuff: The Materials the World is Made of*. New York: Basic Books, 1997.

Attenborough, D. *Life on Earth*. Boston, MA: Little Brown, 1979.

Barrow, J. D. *Impossibility: The Limits of Science and the Science of Limits*. Oxford: Oxford University Press, 1998.

Berg, P. and M. F. Singer *Dealing with Genes: The Language of Heredity*. Sausalito, CA: University Science Books, 1972.

Brun, G., L. McKane, and G. Karp *Biology: Exploring Life*, 2nd edition. New York: Wiley, 1993.

Harre, R. *Great Scientific Experiments*. New York: Oxford University Press, 1987.

Holton, G. *Thematic Origins of Scientific Thought: Kepler to Einstein*, revised edition. Cambridge, MA: Harvard University Press, 1988.

Lehninger, A. L., D. L. Nelson, and M. M. Cox *Principles of Biochemistry*, 2nd Edition. New York: Worth, 1993.

Press, F. and R. Siever *Understanding Earth*, 2nd edition. New York: Freeman, 1997.

Skinner, B. J. and S. C. Porter *The Blue Planet*. New York: Wiley, 1998.

Snyder, C. H. *The Extraordinary Chemistry of Ordinary Things*. New York: Wiley, 1992.

von Baeyer, H. C. *Maxwell's Demon*. New York: Random House, 1998.

Zeilik, M. *Astronomy: The Evolving Universe*, 6th edition. New York: Wiley, 1991.

The following titles provide in depth, but accessible, treatments of more specialized topics.

Amdahl, K. *There Are No Electrons*. Arvada, CO: Clearwater, 1991.

Andrade, N. *Rutherford and the Nature of the Atom*. New York: Doubleday, 1964.

———. *Sir Isaac Newton: His Life and Work*. New York: Doubleday, 1958.

Armbruster, P. and F. Hessberger, "Making New Elements," *Scientific American*, September, 1998, pp. 72–77.

Atkins, P. W. *The Second Law*. New York: Scientific American Library, 1984.

Benedick, R. *Ozone Diplomacy*. Cambridge, MA: Harvard University Press, 1991.

Boss, A. *Searching for Earths: The Race to Find New Solar Systems*. New York: Wiley, 1998.

Burchfield, J. D. *Lord Kelvin and the Age of the Earth*. Chicago, IL: University of Chicago Press, 1990.

Caspar, M. *Kepler 1571–1630*. New York: Collier, 1959.

Cassidy, D. *The Life and Times of Werner Heisenberg*. New York: Freeman, 1992.

Christianson, G. *Edwin Hubble: Mariner of the Nebulae*. New York: Farrar, Straus, and Giroux, 1995.

Cohen, I. B. *The Birth of a New Physics*. New York: Doubleday, 1960.

Darwin, C. *The Origin of Species*. New York: Penguin, 1985 reprint of 1859 edition.

Deamer, D. W. and G. R. Fleischaker *Origin of Life: The Central Concepts*. Boston, MA: Jones and Bartlett, 1994.

DeDuve, C. *Vital Dust: Life as a Cosmic Imperative*. New York: Basic Books, 1995.

Dibner, B. *Alessandro Volta and the Electric Battery*. New York: Franklin Watts, 1964.

Drake, S. *Galileo: Pioneer Scientist*. Toronto, Canada: University of Toronto Press, 1990.

Dudley, W. *Genetic Engineering, Opposing Viewpoints*. San Diego, CA: Greenhaven Press, 1990.

Emsley, J. *The Elements*, 3rd edition. New York: Oxford University Press, 1998.

Everitt, C. W. F. *James Clerk Maxwell: Physicist and Natural Philosopher*. New York: Scribners, 1976.

Ferris, T. *Coming of Age in the Milky Way*. New York: William Morrow, 1988.

Gonick, L. and M. Wheelis *The Cartoon Guide to Genetics*, 2nd edition. New York: HarperCollins, 1991.

Gordon, J. E. *The New Science of Strong Materials*. Princeton, NJ: Princeton University Press, 1976.

Gould, S. J. *Ever Since Darwin*. New York: Norton, 1978. Also other collections of essays published by Norton, including *The Panda's Thumb* (1980), *Hen's Teeth and Horse's Toes* (1983), and *The Flamingo's Smile* (1985).

Greeley, R. *Planetary Landscapes*, 2nd edition. New York: Chapman and Hall, 1993.

Hartmann, W. K. *Moons and Planets*, 3rd edition. Belmont, CA: Wadsworth, 1993.

Hawkings, G. S. *Stonehenge Decoded*. New York: Doubleday, 1965.

Hawking, S. *A Brief History of Time*. New York: Bantam, 1988.

Hazen, R. M. *The Breakthrough: The Race for the Superconductor*. New York: Ballantine, 1988.

———. *The Diamond Makers*. New York: Cambridge University Press, 1999.

———. *The New Alchemists: Breaking the Barriers of High–Pressure Research*. New York: Times Books, 1993.

Herman, R. *Fusion: The Search for Endless Energy*. Cambridge: Cambridge University Press, 1990.

Hoffmann, R. *The Same and Not the Same*. New York: Columbia University Press, 1995.

Horgan, J. *The End of Science*. New York: Times Books, 1997.

Jastrow, R. *Red Giants and White Dwarfs*. New York: Norton, 1990.

Kuhn, T. *The Copernican Revolution*. New York: Random House, 1959.

———. *The Structure of Scientific Revolutions*. Chicago, IL: University of Chicago Press, 1962.

Lawrence, P. *The Making of a Fly: The Genetics of Animal Design*. Oxford: Blackwell, 1992.

Levy, P. *The Periodic Table*. New York: Schocken, 1984.

Lightman, A. *Einstein's Dreams*. New York: Pantheon, 1993.

McPhee, J. *Annals of the Former World*. New York: Farrar, Straus, and Giroux, 1998.

Michaels, P. *The Sound and the Fury: The Science and Politics of Global Warming*. Washington, DC: Cato Institute, 1992.

Morowitz, H. *The Beginnings of Cellular Life*. New Haven, CT: Yale University Press, 1992.

Morris, R. *Cosmic Questions: Galactic Halos, Cold Dark Matter, and the End of Time*. New York: Wiley, 1993.

———. *Time's Arrow: Scientific Attitudes Toward Time*. New York: Touchstone, 1985.

National Academy of Sciences. *Science and Creationism: A View from the National Academy of Sciences*. Washington, DC: National Academy Press, 1984.

———. *Teaching About Evolution and the Nature of Science*. Washington, DC: National Academy Press, 1997.

Pais, A. *Subtle is the Lord: The Science and Life of Albert Einstein*. New York: Oxford University Press, 1982.

Pera, M. *The Ambiguous Frog: The Galvani-Volta Controversy on Animal Electricity*. Princeton, NJ: Princeton University Press, 1992.

Perrin, J. *Atoms*. Woodbridge, CT: Ox Bow Press, 1990 (translation of the 1913 French edition).

Pflaum, R. *Grand Obsession: Madame Curie and Her World*. New York: Doubleday, 1989.

Rhodes, R. *The Making of the Atomic Bomb*. New York: Simon and Schuster, 1986.

Riordan, M. *The Hunting of the Quark*. New York: Simon and Schuster, 1987.

Rudwick, M. *The Meaning of Fossils*, 2nd edition. Chicago, IL: University of Chicago Press, 1976.

Sobel, M. I. *Light*. Chicago, IL: University of Chicago Press, 1989.

Sullivan, W. *We Are Not Alone*, revised edition. New York: Plume, 1994.

Thomson, G. *J. J. Thomson, Discoverer of the Electron*. New York: Doubleday, 1966.

Trefil, J. *From Atoms to Quarks*. New York: Anchor, 1993.

van Spronsen, J. W. *The Periodic System of Chemical Elements*. New York: Elsevier, 1969.

von Baeyer, H. C. *Taming the Atom: The Emergence of the Visible Microworld*. New York: Random House, 1992.

Ward, P. *The End of Evolution: On Mass Extinctions and the Preservation of Biodiversity*. New York: Bantam, 1994.

Watson, J. D. *The Double Helix*, revised edition. New York: Atheneum, 1985.

Weiner, J. *The Beak of the Finch*. New York: Knopf, 1994.

Wilson, E. O. *The Diversity of Life*. New York: Norton, 1992.

Wood, R. M. *The Dark Side of the Earth*. London: Allen and Unwin, 1985.

Notes

Notes